Ernst Schering Research Foundation Workshop 26
Recent Trends in Molecular Recognition

Springer-Verlag Berlin Heidelberg GmbH

Ernst Schering Research Foundation
Workshop 26

Recent Trends
in Molecular Recognition

F. Diederich, H. Künzer
Editors

With 99 Figures and 8 Tables

Springer

Series Editors: G. Stock and U.-F. Habenicht

ISSN 0947-6075

CIP data applied for

Die Deutsche Bibliothek – CIP-Einheitsaufnahme
Schering-Forschungsgesellschaft <Berlin>: Ernst Schering Research Foundation Workshop. - Berlin; Heidelberg; New York; Barcelona; Budapest; Hong Kong; London; Milan; Paris; Santa Clara; Singapore; Tokyo: Springer.
ISSN 0947-6075
26. Recent trends in molecular recognition. - 1998
Recent trends in molecular recognition: with 8 tables / F. Diederich; H. Künzer ed. - Berlin; Heidelberg; New York; Barcelona; Hong Kong; London; Milan; Paris; Singapore; Tokyo: Springer, 1998
(Ernst Schering Research Foundation Workshop; 26)
ISBN 978-3-662-03576-4   ISBN 978-3-662-03574-0 (eBook)
DOI 10.1007/978-3-662-03574-0

© Springer-Verlag Berlin Heidelberg 1998
Originally published by Springer-Verlag Berlin Heidelberg New York in 1998.
Softcover reprint of the hardcover 1st edition 1998

The use of general descriptive names, registered names, trademarks, etc. in this publication does not imply, even in the absence of a specific statement, that such names are exempt from the relevant protective laws and regulations and therefore free for general use. Product liability: The publishers cannot guarantee the accuracy of any information about dosage and application contained in this book. In every individual case the user must check such information by consulting the relevant literature.

Typesetting: Data conversion by Springer-Verlag
Printing: Druckhaus Beltz, Hemsbach. Binding: J. Schäffer GmbH & Co. KG, Grünstadt
SPIN: 10534140   13/3135-5 4 3 2 1 0 – Printed on acid-free paper

Preface

Reasoning in terms of molecular recognition may be traced back to Emil Fischer, who practiced the art of chemistry at Humboldt University in Prussian Berlin a century ago. He was a man of seminal vision when he postulated in 1894 that enzymes recognize their substrates according to the lock-and-key principle. I am quite convinced that he was fully aware of the future impact of this principle, which he enunciated far ahead of its time. It was actually ignored for almost 30 years before being rediscovered by others and given its rightfully important stature and influence. My conviction is based on the unique instinct of Fischer in many other matters; remember that he correctly predicted the absolute configuration of D-glyceraldehyde and thus of the carbohydrates 50 years before the Bijvoet X-ray structure of the sodium-rubidium salt of (+)-tartaric acid was elucidated.

Today, it is clearly recognized that molecular recognition impacts and determines all life processes. It has become a key research field in both chemistry and biology and the emerging interface of what now is being called "chemical biology."

When browsing through the December 1997 issue of *Angew. Chem.*, for example, 22 out of 43 contributions could be classified as dealing with various aspects of supramolecular chemistry and molecular recognition. This is representative for all major journals such as *Chem. Commun.*, *Angew. Chem.*, *J. Am. Chem. Soc.*, and, when including biomolecular recognition, it is to a large extent also true for *Nature* and *Science.*

Molecular recognition research pursues the atomistic understanding of individual intermolecular interactions in molecular complexes and

The Participants of the workshop

large assemblies of chemical or biological small macromolecules. The technological advances derived from this knowledge are particularly important, diverse, and directly evident in the pharmaceutical industry in areas such as:

- Rational, X-ray, or NMR structure-based drug design and lead optimization.
- Identification through genomics, of well-defined enzymes and receptors as future targets for drug discovery. It will be increasingly relevant, given the extremely large number of possible targets opened up by genomics (e.g., more than 1500 protein kinases alone) to rapidly identify "well-behaved" future targets, such as those that bind their substrates in strongly concave/convex relationships and not based on large protein surface interactions.
- Antisense technology and DNA binding and cleavage towards controlling gene expression are ambitious yet equally important targets in the pharmaceutical and biotech industry as is increasingly the search for rational interfering though small molecules with protein-saccharide recognition.

- Biological transport and bioavailability are important processes that depend on intermolecular interactions and the partitioning of small molecules between various biotissues and require enhanced molecular understanding.
- Bulk chemical production will need more efficient, economic and ecologically balanced synthetic methods. Basically all chemical conversions should be made catalytic and, if needed, stereoselective. The development of stereoselective synthesis and asymmetric catalysis increasingly sees principles of molecular recognition applied to the design of more efficient reagents, catalysts, and templates.

Under the auspices of the Ernst Schering Research Foundation, a 2-day workshop held in Berlin in February 1998 entitled "Recent Trends in Molecular Recognition" addressed novel basic developments of potential relevance to drug research efforts in the pharmaceutical industry. Eleven lectures delivered during this event by a multidisciplinary international panel of renowned scholars are documented in this volume. Although a workshop of this size cannot touch upon all the facets of the subject, readers of this monograph will find a balanced coverage of timely research topics in molecular recognition.

F. Diederich
H. Künzer

Table of Contents

1   Designing Transition Metal Complexes
for Molecular Recognition in Synthetic Transformations
B.M. Trost   1

2   REDOR NMR of Biological Solids:
From Protein Binding Sites to Bacterial Cell Walls
J. Schaefer   25

3   Molecular and Dendritic Receptors for Small Biomolecules
F. Diederich   53

4   Molecular Recognition of DNA by Ecteinascidin 743
B.M. Moore II, F.C. Seaman, and L.H. Hurley   81

5   New Tools for Drug Design
Based on Protein Ligand Recognition Principles
G. Klebe 97

6   Sequence-Specific Recognition of DNA
and Control of Gene Expression
by Oligonucleotide-Intercalator Conjugates
C. Hélène, C. Giovannangeli, J.-S. Sun, and T. Garestier . . 119

7   Combinatorial Nucleic Acid Libraries:
The New World of Aptamers and Ribozymes
M. Famulok 135

8   Sequence Specific Recognition of Double Stranded DNA
     by Peptide Nucleic Acid
     P.E. Nielsen 151

9   Solid Phase Libraries of Glycopeptide Templates in the Study
     of Complex Oligosaccharide-Receptor Interactions
     M. Meldal, P.M. StHilaire, and K. Bock 169

10   The Molecular Recognition of Saccharides
     and Glycoprotein-Inspired Materials
     L.L. Kiessling 183

11   Self-Organized Autocatalytic Chemical Networks
     and Molecular Ecosystems:
     Do They Provide the Experimental Tools for Modeling
     the Transition from Inanimate to Animate Chemistry?
     M.R. Ghadiri 213

Subject Index 239

Previous Volumes Published in this Series 245

List of Editors and Contributors

Editors

F. Diederich
Laboratorium für Organische Chemie, Swiss Federal Institute of Technology,
Universitätstrasse 16, 8092 Zürich, Switzerland

H. Künzer
Research Laboratories of Schering AG, Department of Pharmaceutical
Chemistry III, Müllerstrasse 178, 13342 Berlin, Germany

Contributors

K. Bock
Carlsberg Laboratory, Department of Chemistry, 10 Gamle Carlsberg Vej,
2500 Valby, Denmark

F. Diederich
Laboratorium für Organische Chemie, Swiss Federal Institute of Technology,
Universitätstrasse 16, 8092 Zürich, Switzerland

M. Famulok
Institute of Biochemistry, Ludwig Maximilians University,
Feodor-Lynen-Strasse 25, 81377 Munich, Germany

T. Garestier
Biophysics Laboratory, National Natural History Museum, INSERM U 201,
CNRS URA 481, 43, rue Cuvier, 75231 Paris Cédex 05, France

M.R. Ghadiri
Departments of Chemistry and Molecular Biology and The Skaggs Institute
for Chemical Biology, 10550 North Torrey Pines Road, La Jolla, CA 92037,
USA

C. Giovannangeli
Biophysics Laboratory, National Natural History Museum, INSERM U 201,
CNRS URA 481, 43, rue Cuvier, 75231 Paris Cédex 05, France

C. Hélène
Biophysics Laboratory, National Natural History Museum, INSERM U 201,
CNRS URA 481, 43, rue Cuvier, 75231 Paris Cédex 05, France

L.H. Hurley
Drug Dynamics Institute, College of Pharmacy, The University of Texas
at Austin, Austin, TX 78712-1074, USA

G. Klebe
Institute of Pharmaceutical Chemistry, Philips-University of Marburg,
Marbacher Weg 6, 35032 Marburg, Germany

L.L. Kiessling
Departments of Chemistry and Biochemistry, University of Wisconsin,
1101 University Avenue, Madison, WI 53706, USA

M. Meldal
Carlsberg Laboratory, Department of Chemistry, 10 Gamle Carlsberg Vej,
2500 Valby, Denmark

B.M. Moore II
Drug Dynamics Institute, College of Pharmacy, The University of Texas
at Austin, Austin, TX 78712-1074, USA

P.E. Nielsen
Center for Biomolecular Recognition, The Panum Institute, Blegdamsvej 3c,
2200, Copenhagen N, Denmark

J. Schaefer
Department of Chemistry, Washington University, One Brookings Drive,
St. Louis, MO 63130-4899, USA

F.C. Seaman
Drug Dynamics Institute, College of Pharmacy, The University of Texas
at Austin, Austin, TX 78712-1074, USA

P.M. StHilaire
Carlsberg Laboratory, Department of Chemistry, Gamle Carlsberg Vej 10,
2500 Valby, Denmark

J.-S. Sun
Biophysics Laboratory, National Natural History Museum, INSERM U 201,
CNRS URA 481, 43, rue Cuvier, 75231 Paris Cédex 05, France

B.M. Trost
Department of Chemistry, Stanford University, Stanford, CA 94305-5080,
USA

1 Designing Transition Metal Complexes for Molecular Recognition in Synthetic Transformations

B.M. Trost

1.1   Introduction 1
1.2   Asymmetric Desymmetrization 7
1.3   Conformationally Constrained Ligands 10
1.4   Asymmetric Induction at the Pronucleophile 14
1.5   Conclusions 23
References 23

1.1 Introduction

While the concept of molecular chirality celebrated its 100th birthday some years ago, the impact of this fundamental structural issue has been slow to gain prominence in synthetic chemistry. Except for the obvious situation wherein enantiomerically pure starting materials, largely from nature, were the synthetic building blocks to permit access to enantiomerically pure products, the difficulties of controlling relative stereochemistry made the more difficult task of controlling absolute stereochemistry out of reach. The realization of the potential profoundly different effects that enantiomers can have in biological applications dictated that chiral products should be introduced solely as the enantiomerically beneficial isomer with few exceptions. The challenge to invent synthetic protocols to generate enantiomerically pure products was presented to the synthetic community.

A number of strategies for meeting this objective have evolved. A conceptually attractive one is the employment of asymmetric catalysis – both biological and abiological. While each will have their roles to play, the almost unlimited types of non-biological chemical reactions that are available or can be invented makes imparting catalytic enantioselectivity to them a highly desirable goal.

Alkylation reactions in contrast to hydrogenations (which form C-H bonds) or hydroxylations/epoxidations (which form C-O bonds) can form many different types of bonds to carbon, including but not limited to C-C, C-N, C-S, C-P, as well as C-H and C-O. Metal catalyzed versions of such reactions then become candidates for asymmetric induction. While metal catalyzed allylic alkylations constitute just one type of alkylation, its diversity makes it an exciting target and its complexity a difficult one (Trost and Van Vranken 1996). Scheme 1 illustrates a catalytic cycle. The first issue derives from the fact that both reacting partners, the nucleophile and the allyl unit, may be the site for asymmetric induction. Typical carbon nucleophiles would present pro-chiral faces, which could be differentiated in step 3, the addition. Introduction of asymmetry into the allyl fragment can occur at steps 1, 2, or 3, i.e., in the initial complexation, ionization, or addition. The issue is further complicated by the fact that the ionization and/or addition steps may occur with either retention or inversion of configuration at the carbon undergoing substitution.

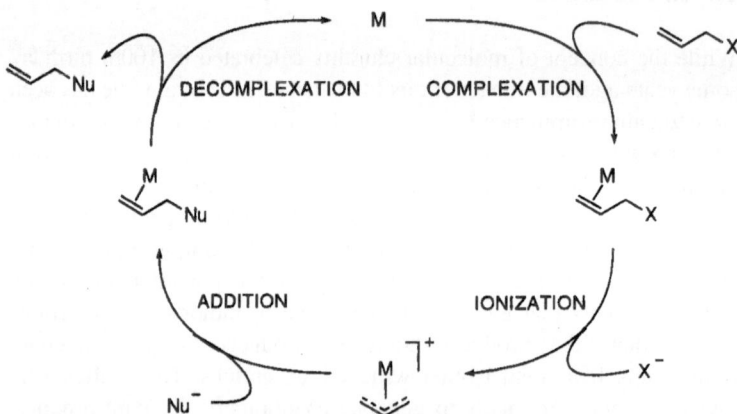

Scheme 1. A catalytic cycle for allylic alkylation

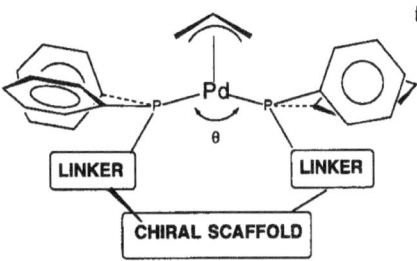

Fig. 1. A structural model

In palladium catalyzed reactions, ionizations and nucleophilic additions occur outside the coordination sphere of the metal. Since the chiral information resides in ligands attached to the metal, the prime issue becomes how transfer of this information can occur across the barrier of the allyl unit in a square planar complex as exists for the palladium reaction. While a number of models evolved to address this difficulty, our approach borrowed some fundamental concepts from enzymes (Trost and Murphy 1985; Trost and Van Vranken 1992; Trost et al. 1992). As depicted in Fig. 1, primary chirality that resides in a scaffold will create conformational chirality, which in turn creates chiral space. By the chiral space serving as a molecular receptor, the substrate will reside in this "active site." The need for the allyl precursor to enter the active site could induce asymmetric induction in the complexation step. Alternatively, both the ionization and addition steps require molecular motions wherein η^2 and η^3 complexes are involved. The ability of the chiral space to influence the molecular motions, as illustrated in Fig. 2, would influence enantiodiscrimination in either an ionization or addition event in complementary fashion, i.e., ionization involves an η^2 complex going to η^3 and addition the reverse. Thus, in II (Fig. 2), as viewed from the top, substrate undergoes a counterclockwise motion in the ionization pocket in the case of the R,R ligand of complex I in the enantiodiscriminating step and in III a clockwise motion.

Using this model, three areas for variation have been explored, the chiral scaffold, the linkers, and the diarylphosphino binding posts. Figure 3 illustrates some of the modules that have been examined and Fig. 4 some of the ways the modules have been assembled to create specific ligands.

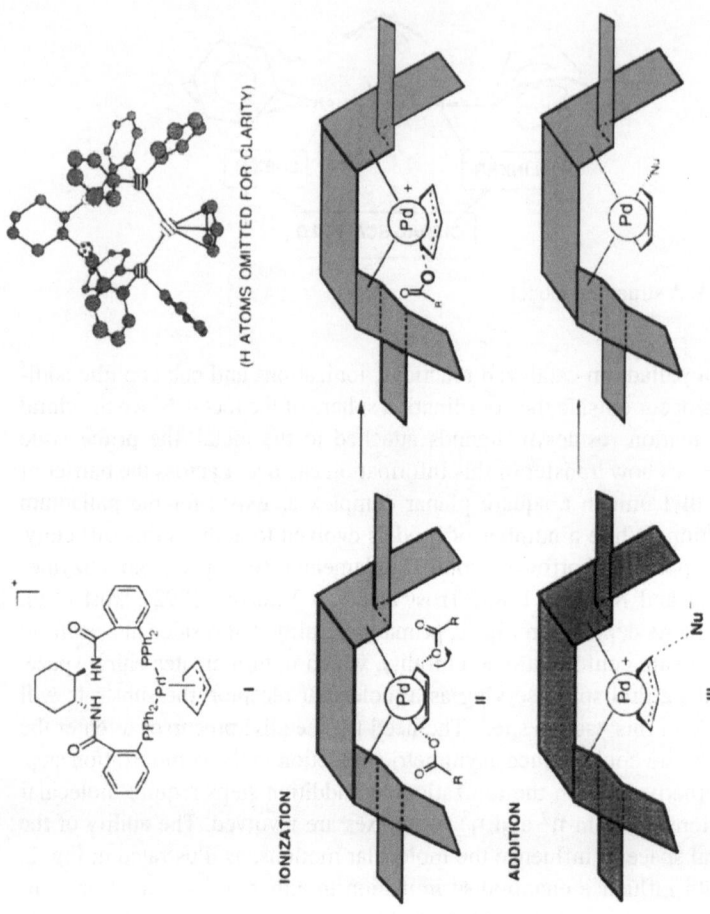

Fig. 2. A functional model for an asymmetric catalyst

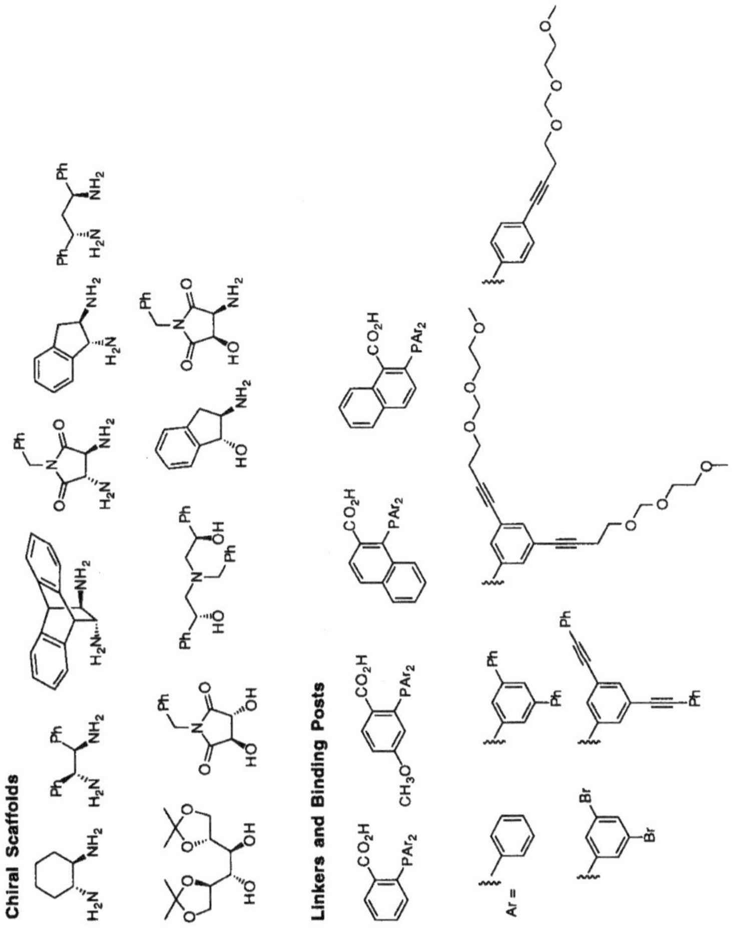

Fig. 3. Structural modules for chiral ligands

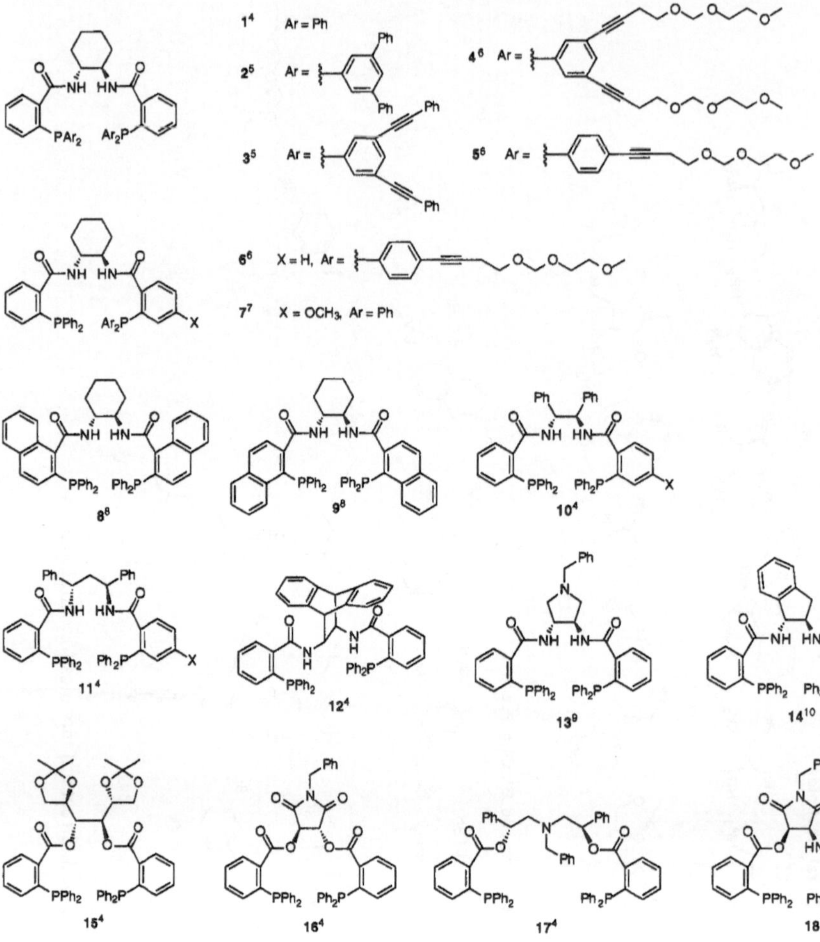

Fig. 4. Example of chiral ligands

1.2 Asymmetric Desymmetrization

Asymmetric induction in the ionization step is uncomplicated by issues arising from involvement of the nucleophile in the enantiodiscriminating step. Thus, desymmetrizations of a meso-2-ene-1,4-diol as depicted in Eqs. 1 and 2 were used as test reactions. Reactions employing the ester derived ligands **15–17** gave ee's on the order of 60–65%. In contrast, the tighter amide type ligands **1** and **11–13** enhanced the ee to 80–90% (Trost et al. 1992). The simple addition of triethylamine in the latter cases further increased the ee's to 95% (D. Patterson, unpublished). From this data, it is clear that the separation of the diol or diamine in terms of the functionality being 1,2 vs 1,3 vs 1,4 or even vs 1,5 on the chiral scaffold has little effect on the enantiodiscrimination.

A most interesting effect arose from consideration of a series of tartrate derived ligands **16, 18, 19,** and **20** (Trost et al. 1994). By

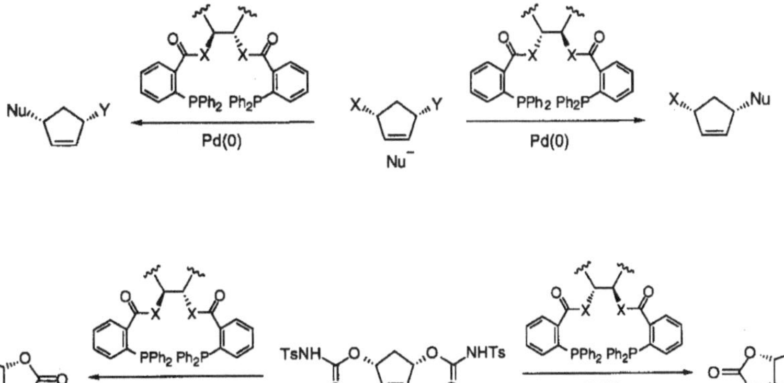

(1)

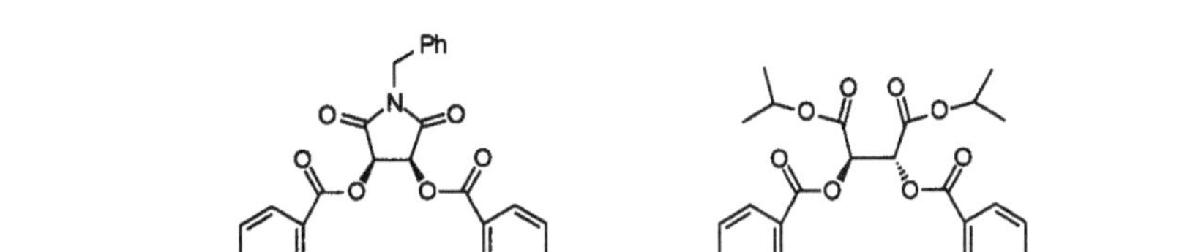

(2)

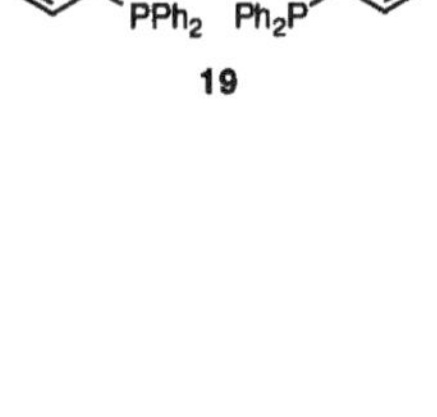

19

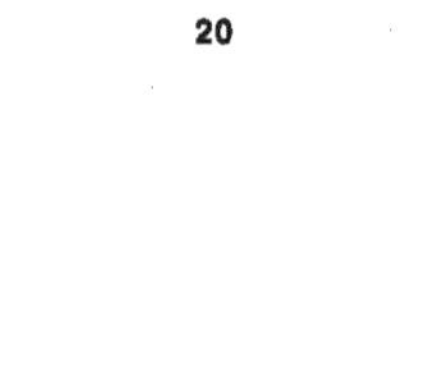

20

definition, the achiral ligand **19** must give rise to racemic product. Considering the closeness in structure of an ester linkage and a secondary amide in terms of their conformations, the desymmetrized ligand **18** would be expected to resemble the meso ligand **19** and might be considered to be a pseudo-meso object. It is quite surprising, then, that ligand **18** gives an ee (78%) in reaction 2 that is even somewhat higher than that seen with the normal chiral tartrate ligand **16** (75% ee) and over twice as high as observed for the chiral object **20**.

How can a pseudo-meso object give an ee higher than a chiral object? An answer may derive from consideration of why a meso object like **19** gives rise to a racemic product. Based upon our model of Fig. 1, the asymmetric induction derives from chiral space created by conformational chirality derived from the two diphenylphosphino moieties. This chiral space exists regardless of whether the scaffold is chiral or achiral. An achiral scaffold gives rise to racemic products because the two enantiomeric chiral spaces are equal in energy. Thus, reaction occurs equivalently in both. The chiral scaffold induces an energy difference between the two chiral spaces and thus an asymmetric induction. Simple molecular mechanics suggests that the energy difference between the two chiral spaces derived from ligand **18** is on the order of 5 kcal/mol, more than enough to account for the 78% ee observed.

Another interesting effect derived from the study of the azide displacement is shown in Eq. 3 (Trost et al. 1995). In contrast to the reactions with achiral ligands, which require hours at room temperature,

$$
PhCO_{\prime\prime\prime} \underset{}{\overset{O}{\parallel}} OCPh \quad + \quad TMSN_3 \quad \xrightarrow[{[\eta^3\text{-}C_3H_5PdCl]_2}]{1} \quad PhCO_{\prime\prime\prime} \underset{}{\overset{O}{\parallel}} N_3 \qquad (3)
$$

the asymmetric version was complete in less time than it took to obtain the first tlc (<10 min) at −78°C. Thus, like enzymes, the reaction cavity effects both chiral recognition and rate enhancements. Molecular modeling suggests an explanation for the rate enhancement since, as Fig. 5 shows, substrate binding to the palladium forces a conformation whereby the dihedral angle between the metal and the leaving group is 179°C, practically the ideal stereoelectronic angle for ionization. Thus,

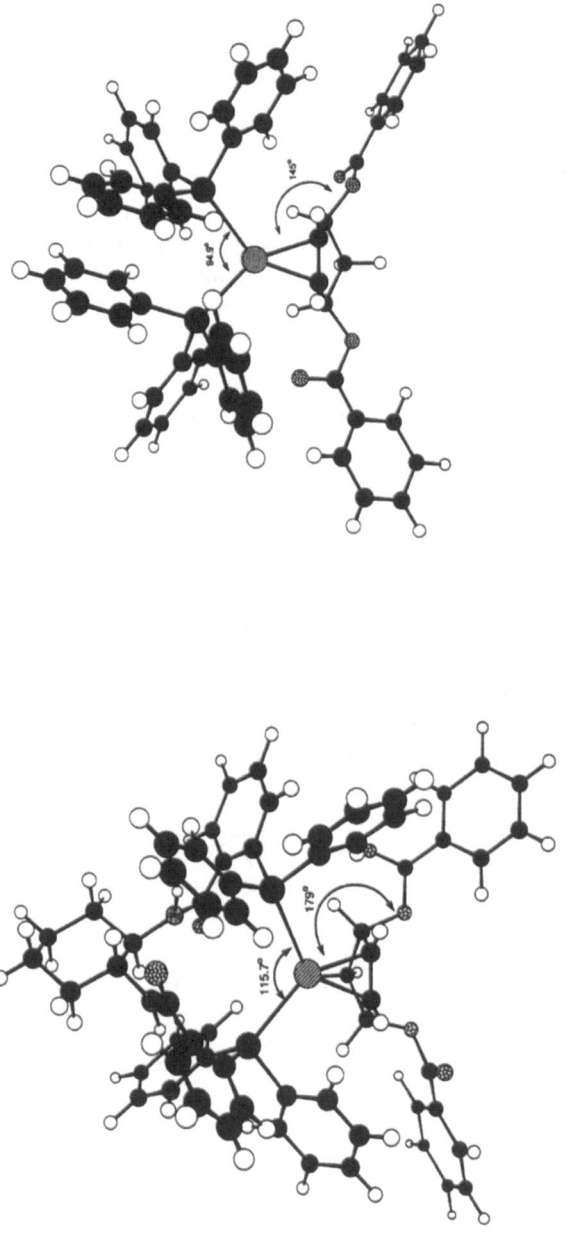

Fig. 5. Substrate binding in chiral and achiral complexes. *Left* with 1 as ligand; *right* with Ph₃P as ligand

part of the energy required for ionization is paid in a pre-equilibrium step, which then requires a lower energy for the transition state of the rate determining bond cleavage step. The same molecular mechanics calculation for the complex formed with triphenylphosphine as ligand shows this dihedral angle to be 145° – far from the ideal. Thus, a considerably higher energetic price must yet be paid for achieving the stereoelectronics required for ionization in the rate determining bond cleavage step – thus, a slower reaction. Imposing a conformation for binding that is more reactive in the subsequent rate determining step is a phenomenon that has been used to explain the rate accelerations of numerous enzymatic reactions.

1.3 Conformationally Constrained Ligands

Reactions of vinyl epoxides for asymmetric induction offer a special challenge for ligand design. As shown in Scheme 2, two major selectivity issues arise – regio- and enantioselectivity. In fact, previous results suggest that 1,4-addition, not 1,2-addition, should dominate. Thus, for asymmetric induction, we require the catalyst to override the normal regioselectivity biases of the system. Since the substrate is racemic, the reaction requires a racemization event to avoid simply being a kinetic resolution. The dynamics of π-allylpalladium complexes permit just such a racemization by a unimolecular η^3-η^1-η^3 mechanism or a bimolecular palladium-palladium substitution.

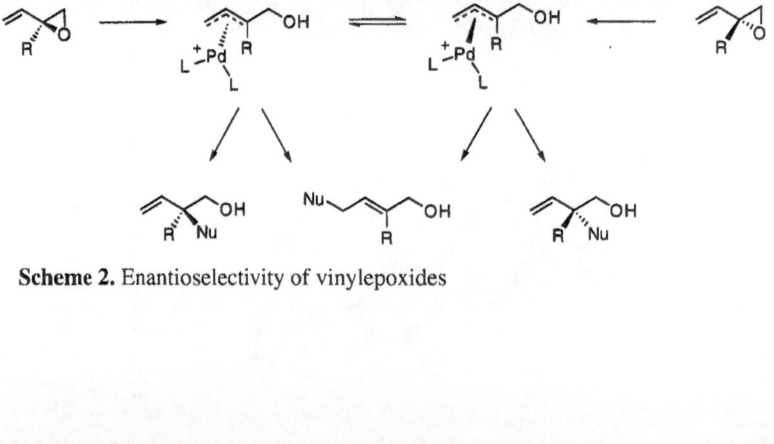

Scheme 2. Enantioselectivity of vinylepoxides

$$(4)$$

The reaction of butadiene monoepoxide with phthalimide utilizing our "standard" ligand **1** gratifyingly showed a 10:1 regioselectivity for the internal attack product **21** (Eq. 4; Trost and Bunt 1996). The highest ee that could be achieved, however, was 76%.

Assessing that the limitation in ee could lay in an unfavorable rate of racemization of the π-allyl intermediates compared to the rate of nucleophilic attack, ligand modification was viewed to be an option. A more constrained reaction pocket was viewed as one way to slow the rate of nucleophilic attack and thereby enhance the ee. The introduction of peri interactions by switching the linker to a naphthoic acid rather than a benzoic acid would restrict the conformations of the diphenylphosphino moiety as in **9** and the amide moiety as in **8**. Interestingly, ligand **9** led to low asymmetric induction. This observation further supports the working model regarding the importance of the conformations of the two diphenylphosphino moieties. In this case, they apparently are constrained from creating effective chiral space. On the other hand, constraining the amide bond dramatically increased the **21:22** ratio to better than **75:1** but also saw the ee jump to 98% (Trost and Bunt 1996). Similar results are obtained with isoprene monoepoxide (T.L. Calkin, unpublished).

Use of a chiral nucleophile raises the question of whether the reaction will be catalyst or substrate controlled. Using natural amino acids, good regioselectivity for attack at the internal position is observed for isoprene monoepoxide as shown in Eqs. 5 and 6. Good complementary diastereoselectivity is observed with the naphthyl based ligands as shown (C. Oertelt and T.L. Calkins, unpublished). Even with the standard benzo-type ligand **10** and its enantiomer, good catalyst controlled selectivity was observed (**10** gave **23:24** of **12:1**, **ent-10** gave **23:24** of **1:14**) but in somewhat lower yields (45–50%).

The synthesis of vinylglycinol in one step from two readily available starting materials constitutes the most practical synthesis to date of this important building block. Several simple transformations illustrate the utility. For example, oxidative cleavage of **25**, derived by exchanging the phthalimido group for acetamido, provides access to R-serine (Eq. 7; R.C. Lemoine, unpublished). Schemes 3 and 4 illustrate facile syntheses of clinically important agents, vigabatrin for epilepsy (Trost and Lemoine 1996) and ethambutol for tuberculosis (R.C. Lemoine, unpublished). Conversion of the hydroxyl group of **21** into a triflate sets the stage for a malonate displacement and hydrolysis to generate R-vigabatrin. Thus, this becomes available in only four steps in 59% yield. It should be self-evident that S-vigabatrin, the enantiomer that is the pharmaceutically important agent, should also be available in a similar overall yield simply by inverting the ligand stereochemistry in the first step.

Equation 8 shows the easy access to S-2-amino-1-butanol from phthalimide **21**. This constitutes a formal synthesis since previously alkylation with 1,2-dichloroethane led to (+)-ethambutol. The requirement for excess of the valuable chiral amine and low yield in the alkylation led us to develop an alternative route shown in Scheme 4.

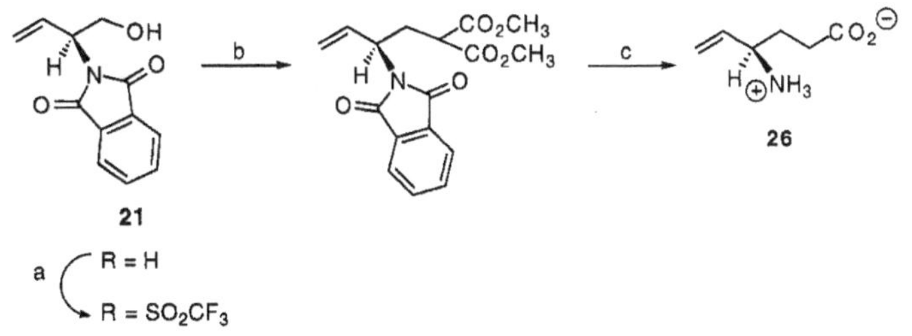

a) (CF$_3$SO$_2$)$_2$O, (C$_2$H$_5$)$_3$N, CH$_2$Cl$_2$, 0°, 96% yield. b) CH$_2$(CO$_2$CH$_3$)$_2$, NaH, THF, 0°, 64% yield. c) aq. HCl, 100°, 96% yield.

Scheme 3. An asymmetric synthesis of ent-vigabatrin

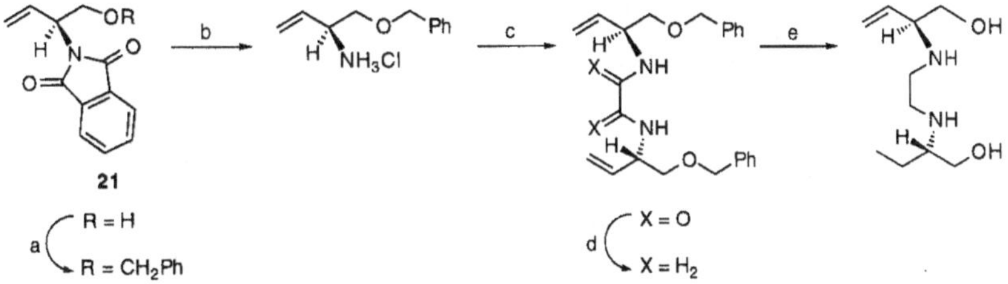

a) NaH, PhCH$_2$Br, DMF, 0°, 82%. b) NH$_2$CH$_2$CH$_2$NH$_2$, C$_2$H$_5$OH, reflux, then HCl, 94%. c) (COCl)$_2$, C$_5$H$_5$N, CH$_2$Cl$_2$, 0°, 97%. d) Red-Al, PhCH$_3$, 45°, 78%. e) H$_2$, Pd/C, CH$_3$OH then H$_2$, Pd/C, HCl, CH$_3$OH, 74%.

Scheme 4. An asymmetric synthesis of ethambutol

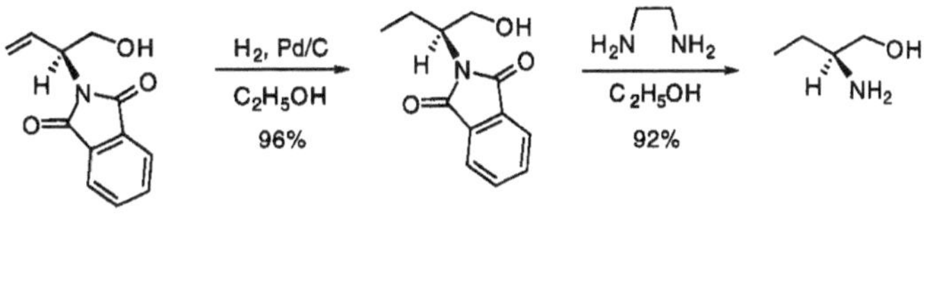

(8)

Protection of the primary alcohol was required for the successful elaboration of the ethano bridge. All attempts at direct reductive alkylation
with glyoxyal or its equivalents failed. On the other hand, the oxalamide
was readily reduced with red-Al and the synthesis completed by double
bond reduction and hydrogenolysis over palladium on carbon. While the
synthesis from butadiene monoepoxide and phthalimide via the direct
alkylation of S-2-amino-1-butanol is only four steps, the actual yield
based upon the aminoalcohol is difficult to evaluate because of the
requirement of excess amine in the final alkylation. On the other hand,
the route of Scheme 4 requires six steps and proceeds in a satisfactory
42% overall yield from the same building blocks.

1.4 Asymmetric Induction at the Pronucleophile

Consideration of whether the asymmetric induction can be extended to
a pronucleophile as suggested in Fig. 6 raises the question of the dimensions of the "chiral space." As depicted, the newly created stereogenic
center in such a case is even more distant. Thus, it was not surprising
that one of the very early studies, which involved just an asymmetric
induction, failed. As a result, an alternative strategy took cognizance of
the fact that charged pro-nucleophiles have counter ions. If cationic
binding sites are built proximally to the "chiral space," they could have
two beneficial effects: (1) increase the depth of the "chiral space" and

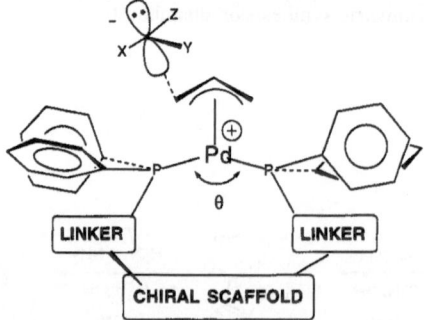

Fig. 6. Model for asymmetric induction at pronucleophile

(2) provide orientation of the attacking nucleophile. Figure 4 depicts a number of such systems wherein the metal binding sites are simple polyethers. In ligand **4**, each aryl ring of the diarylphosphino moiety has two such "arms" in the 3,5-positions for a total of eight. In a π-allylpalladium complex derived from ligand **4**, the flexible "arms" stretched out in solution may bind a cation via coordination with the ether oxygens. By folding, they could help deliver the cation and its accompanying anion to the complexed π-allyl unit to generate the desired product. Modeling suggests a resemblance of this eight-armed complex to an octopus, which led to their designation as "octopus ligands." Unfortunately, the eight armed "creature" proved unreactive. Apparently, the opening to the "chiral space" was blocked by too many arms.

Figure 7 depicts ligand **5** complexed to palladium bearing a π-cyclopentenyl unit. It is perhaps more apparent how binding of one arm to the cation of the ion pair may then help to deliver the nucleophile to the

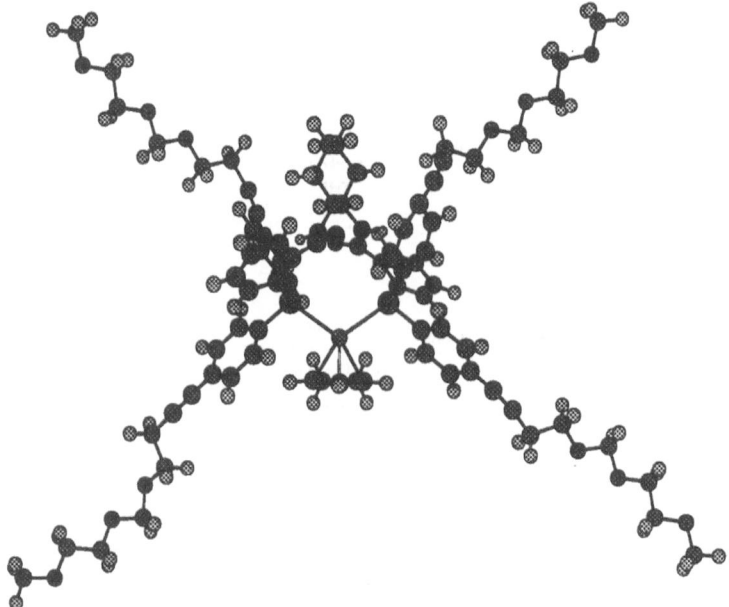

Fig. 7. A cyclopentenyl complex with tetrapodal ligand **5**

π-allyl unit. In order to assess whether such designed ligands did play a role, the reactions of racemic 3-methoxycarboxycyclohexene with dibenzyl malonate were examined (Eq. 9; Trost and Radinov 1997). Interestingly, the reaction proceeded at room temperature within 2 h to give a quantitative yield of alkylated product of 99% ee. A significant rate enhancement was observed since the "standard" ligand **1** required 16 h to give similar results (94% yield, 98% ee).

The dipodal ligand **6** appears to lead to faster rates than the tetrapodal ligand **5**. For example, the difficult acyclic 1,3-dimethyl substrate required 40°C for reaction to occur with the palladium complex of the tetrapodal ligand **5** and thereby a lowering of the ee to 76% (Eq. 10). On the other hand, the dipodal ligand **6** allowed reaction to proceed at ambient temperature, which resulted in a 90% ee.

Reactions with heteroatom nucleophiles such as sodium phthalimide, sodium benzoate, and sodium benzenesulfinate showed similar effects. For example, the reaction of Eq. 11 using the "standard" ligand **1** required tetrahexylammonium bromide as a phase transfer catalyst and required 18 h with 5 mol% palladium precatalyst. On the other hand, the dipod ligand **6**, without any phase transfer catalyst, went to completion

$$\text{(9)}$$

$$\text{(10)}$$

$$\text{(11)}$$

in only 2 h with 0.25 mol% of palladium precatalyst to give the alkylated phthalimide of 98% ee in 91% yield. The tetrapodal ligand **5** also works well with these heteroatom nucleophiles as summarized in Eq. 12 for the oxygen (R. Radinov, unpublished) and sulfur (Trost and Radinov 1997) cases.

$$\text{(structure)} + \text{Na}^{\oplus}\ X^{\ominus} \xrightarrow[\text{CH}_2\text{Cl}_2]{\textbf{5}} \text{(structure)} \qquad (12)$$

$$X = \overset{\ominus}{\text{O}_2\text{CPh}} \qquad 0.5\% \ [\eta^3\text{-}C_3H_5PdCl]_2 \qquad 99\% \quad (90\% \text{ ee})$$

$$X = \overset{\ominus}{\text{O}_2\text{SPh}} \qquad 0.2\% \ (dba)_3Pd_2 \qquad 88\% \quad (99\% \text{ ee})$$

$$\text{(structure, CO}_2\text{CH}_2\text{Ph)} + \overset{R}{\text{(structure, OAc)}} \xrightarrow[\substack{\text{NH} \\ (CH_3)_2N \quad N(CH_3)_2}]{\substack{0.9\% \ \textbf{1} \\ 0.4\% \ [h^3\text{-}C_3H_5PdCl]_2 \\ \text{PhCH}_3, 0°}} \text{(structure, CO}_2\text{CH}_2\text{Ph, R)} \qquad (13)$$

$$\begin{array}{lll} R = H & 94\% & (89\% \text{ ee}) \\ R = CH_3 & 81\% & (95\% \text{ ee}) \\ R = CH_2OAc & 80\% & (94\% \text{ ee}) \end{array}$$

The rate effect was examined more quantitatively for the reaction of Eq. 12, $X = {}^-SO_2Ph$, except that the dipod ligand was employed (Trost and Radinov 1997). For these experiments, the standard reaction parameters were 0.13 mol% $(dba)_3Pd_2 \cdot CHCl_3$ and 0.3 mol% **6** in a water-methylene chloride two-phase mixture at 0°C. Figure 8 involves the sodium salt and dramatically reveals the rate enhancement. Whereas the reaction employing the standard ligand **1** proceeded to less than 10%, after 50 min; the reaction employing the dipod ligand **6** nearly went to completion. If the rate effects derive from metal binding, variation of the metal ion should also have a significant rate effect. Figure 9 illustrates the large rate differences, which is reflected in the time required to reach 50% completion: NH_4^+ (>180 min)<K^+ (180 min)<Li^+ (70 min)<Na^+ (<10 min). The effectiveness of these ligands is particularly noteworthy,

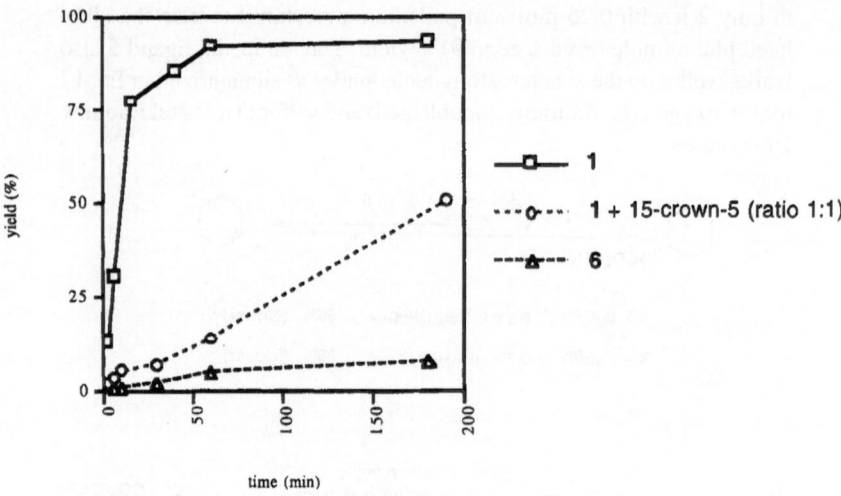

Fig. 8. Rate comparison between standard and dipod ligands

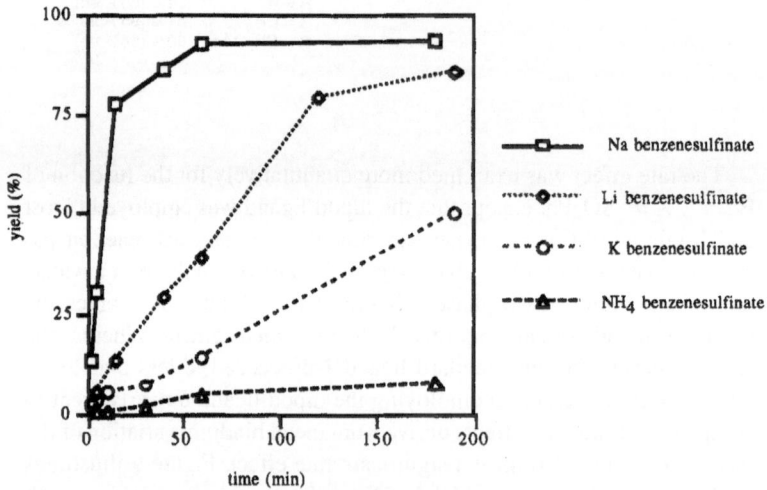

Fig. 9. Cation effect with dipodal ligand **6**

considering that each arm contains only three oxygens for metal coordination in a conformationally unrestricted fashion. The more highly organized crown or cryptand type metal coordinators are not required for significant effects.

Having established the effectiveness of the ligands bearing metal binding arms to direct the nucleophile, we turned to the question of the asymmetric induction at the pro-nucleophile. The initial study began with the "standard" ligand **1**. Employing 2-benzyloxycarbonyl-1-tetralone and allyl acetate, a surprisingly high ee (89%) was already obtained using tetramethylguanidine as base (Trost et al. 1997). Placing substituents at the central carbon increased the ee. The 1,3-disubstituted allylic systems raise the issue of diastereo- as well as enantioselectivity. Gratifyingly, the 1,3-dimethylallyl system participates to give excellent de and ee (Eq. 14). This example undoubtedly involves a syn,syn-1,3-dimethylallylpalladium complex. A cyclic substrate was examined to probe the effect of having an anti,anti geometry in the intermediate. As shown in Eq. 15, the results were even better.

The creation of a quaternary center bearing three different functionalities asymmetrically provides great opportunities for further elaboration. A simple illustration is the asymmetric synthesis of the spiro-alkaloid nitramine, as representative of a class of alkaloids possessing interesting CNS activity. For this synthesis, 2-carboethoxycyclohexanone was alkylated quantitatively under the above conditions to form its 2-allyl derivative of 86% ee

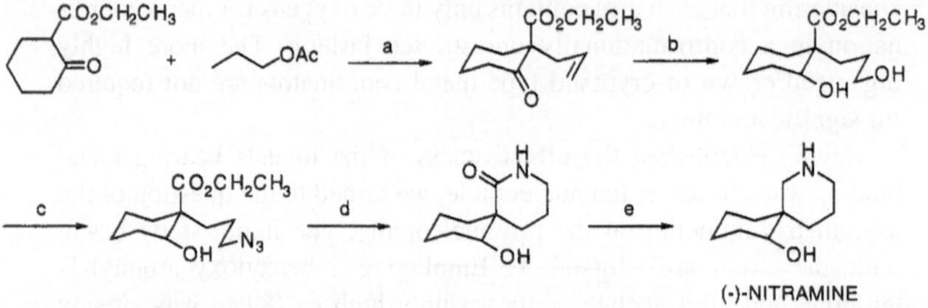

a) As in eq. 13, quantitative yield. b) Disiamylborane, THF, -15° to room temperature, then NaBO$_3$•4H$_2$O, 80%. c) i. (C$_2$H$_5$)$_3$N, CH$_3$SO$_2$Cl, CH$_2$Cl$_2$, -60° to 0°; ii. NaN$_3$, DMF, 40°, 72%. d) H$_2$, Pd(OH)$_2$/C, K$_2$CO$_3$, C$_2$H$_5$OH, room temperature to reflux, 83%. e) LAH, THF, room temperature, 90%.

Scheme 5. An asymmetric synthesis of (-)-nitramine

(Scheme 5). The hydroboration of the alkene effected simultaneous diastereoselective reduction of the ketone to give a single diol, which was elaborated straightforwardly to the target. The spiro alkaloid is available diastereo- and enantiomerically pure in six steps and 43% overall yield from 2-carboethoxycyclohexanone and allyl acetate.

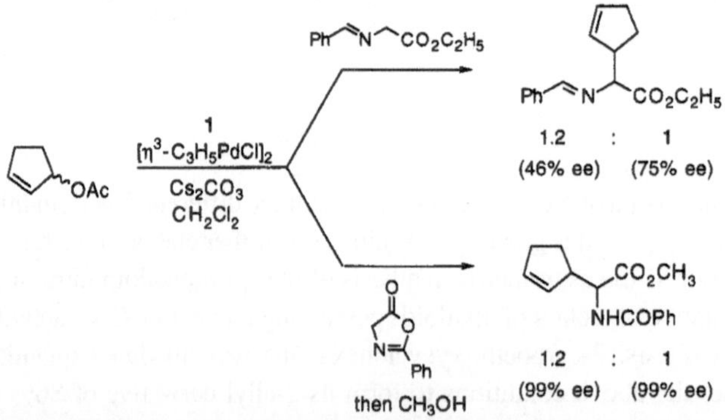

Utilization of this methodology for asymmetric amino acid synthesis required definition of a suitable nucleophile. As shown in Eq. 16, ethyl N-benzylideneglycinate participated to give a nearly 1:1 mixture of diastereomers as expected (X. Ariza, unpublished). Under the conditions of these reactions wherein proton exchange is rapid, the epi-merizable proton α to the ester expectedly rapidly produces the thermo-dynamic mixture at that center. The significant point was the ee. Both showed significant enantioselectivity (46–75%), which was a promising start. Anticipating a conformationally constrained nucleophile would behave better, an azlactone was examined. As expected, the two dias-tereomers again formed in nearly a 1:1 mixture. However, both dias-tereomers were enantiomerically pure.

Using a substituted azlactone precludes α-epimerization in the prod-uct but raises the issue of diastereoselectivity. As shown in Eq. 18, using triethylamine as base and acetonitrile as solvent, good dr as well as excellent ee was obtained (Trost and Ariza 1998). As the size of the R group increased, the diastereoselectivity also increased. The major dias-tereomer was enantiomerically pure in each case.

R	Yield	dr (ee)
CH_3*	90%	8.7 (99%):1(96%)
$PhCH_2$	74%	12.4 (99%):1(96%)
$(CH_3)_2CHCH_2$	77%	13.3 (99%):1(95%)
$(CH_3)_2CH$	87%	>19 (99%):1(-)

*RT

Acyclic allyl systems require substitution on one terminus for satis-factory results (X. Ariza, unpublished). Thus, allylation as in Eq. 18 of 4-benzyl-2-phenylazlactone gave a quantitative yield but the ee of the allylated azlactone was only 40%. On the other hand, alkylations with cinnamyl acetate (Eq. 19) and t-butyl-3-methyl-2-buten-1-yl carbonate (Eq. 20) gave excellent enantioselectivities.

Serine analogues could be accessed via asymmetric desymmetriza-tion of gem-diacetates as illustrated in Eq. 21 (Trost and Ariza 1998). In this case, the best conditions to date employ sodium hydride as base and

$$\text{(19)} \qquad 91\% \ ee$$

$$\text{(20)} \qquad 98\% \ ee$$

$$\text{(21)}$$

R	Yield	dr (ee)
CH₃[*]	74%	6.6 (99%):1(96%)
PhCH₂	81%	9.7 (99%):1(96%)
(CH₃)₂CHCH₂	97%	15 (99%):1(95%)
(CH₃)₂CH	92%	>19 (99%):1(-)

[*]RT

DME as solvent. Exactly the same trends were observed. Once again, the major diastereomers were enantiomerically pure. This method provides a very facile asymmetric access to α-alkyl amino acids, which are difficult to synthesize by other methods. For example, asymmetric hydrogenation methods do not apply. Self-reproduction of chirality cannot access serine derivatives and many of the types of alkyl side chains easily introduced here.

1.5 Conclusions

The need for catalysts for asymmetric synthesis is growing at a phe-nomenal rate. Two broad avenues offer opportunities – biological and abiological. Both will have important roles to play. A major advantage of abiological catalysis is the possibility to semi-rationally design the receptor and to vary its structure systematically for particular reactions. Various structural motifs may be envisioned for asymmetric catalysis. Exploiting the concept of chiral space appears to be a useful paradigm for those transition metal reactions involving enantiodiscriminating steps wherein the bonds are formed or broken outside the coordination sphere of the metal. The resultant catalysts start behaving like enzymes in terms of rate enhancements accompanying asymmetric induction. The success provides strong encouragement for these principles to guide catalyst design in other cases.

Acknowledgement. First and foremost, I am indebted to an exceptional group of coworkers who are individually identified in the references. Financial sup-port was generously provided over the years by the National Science Founda-tion, the National Institutes of Health (General Medical Sciences), and Merck.

References

Trost BM, Ariza X (1997) Catalytic asymmetric alkylation of nucleophiles: asymmetric synthesis of alpha-alkylated amino acids. Angew Chem Int Ed Engl 36:2635

Trost BM, Bunt RC (1996) On ligand design for catalytic outer-sphere reac-tions. A simple asymmetric synthesis of vinylglycinol. Angew Chem Int Ed Engl 35:99–102

Trost BM, Lemoine RC (1996) An asymmetric synthesis of vigabatrin. Tetra-hedron Lett 37:9161–9164

Trost BM, Marschner C (1997) Crafting chiral space. The synthesis of C_2-symmetric diphosphine ligands for an outer-sphere catalytic reaction. Bull Soc Chim France 134:263–274

Trost BM, Murphy DJ (1985) A model for metal-templated catalytic asymmet-ric induction via π-allyl fragments. Organometallics 4:1143–1145

Trost BM, Radinov R (1997) On the effect of a cation binding site in an asym-metric ligand for a catalyzed nucleophilic substitution reaction. J Am Chem Soc 119:5962–9563

Trost BM, Van Vranken DL (1992) Asymmetric ligands for transition metal-catalyzed reactions. 2-Diphenylphosphinobenzoyl derivatives of C_2-symmetrical diols and diamines. Angew Chem Int Ed Engl 31:228–230

Trost BM, Van Vranken DL (1996) Asymmetric transition metal-catalyzed allylic alkylation. Chem Rev 96:395–422

Trost BM, Van Vranken DL, Bingel C (1992) A modular approach for ligand design for asymmetric allylic alkylations via enantioselective palladium-catalyzed ionization. J Am Chem Soc 114:9327–9342

Trost BM, Breit B, Organ MG (1994) On the nature of the asymmetric induction in a palladium catalyzed allylic alkylation. Tetrahedron Lett 35:5817–5820

Trost BM, Stenkamp D, Pulley SR (1995) An enantioselective syntheses of cis-4-tert-butoxycarbamoyl-1-methoxycarbonyl-2-cyclopentene – a useful, general building block. Chem Eur J 1:568–572

Trost BM, Radinov R, Grenzer EM (1997) Asymmetric alkylation of β-ketoesters. J Am Chem Soc 119:7879–7880

2 REDOR NMR of Biological Solids: From Protein Binding Sites to Bacterial Cell Walls

J. Schaefer

2.1	Introduction	26
2.2	Averaging of Dipolar Couplings	26
2.3	Recoupling for Heteronuclear Spin Interactions	27
2.4	Clusters of Labels	28
2.5	EF-Tu	29
2.6	GlnBP	33
2.7	EPSP Synthase	35
2.7.1	Ternary Complex	35
2.7.2	Labeling Strategy	35
2.7.3	^{15}N-Lysine Labeling	36
2.7.4	Residues in the Binding Site	38
2.7.5	^{19}F-Tryptophan Labeling	40
2.8	*Staphylococcus aureus*	42
2.8.1	Bacterial Cell-Wall Peptidoglycan	42
2.8.2	REDOR of the Pentaglycyl Bridge	44
2.8.3	Whole Cells	45
2.8.4	Vancomycin-Resistant Staphylococci	47
2.8.5	Working Model of Glycopeptide Mode of Action	47
2.8.6	Solid-State NMR Detection of Cell-Wall Complexes	48
References		49

2.1 Introduction

Details of the structure and dynamics of proteins, nucleic acids, and their complexes are commonly obtained from two sources: X-ray crystallography and solution-state nuclear magnetic resonance (NMR). However, some proteins and protein complexes crystallize poorly or not at all. Many of these same proteins are insoluble, or aggregate in solution, or exceed the effective molecular-weight limit for solution-state NMR. Such systems may be suitable for analysis by solid-state NMR (Griffiths and Griffin 1993; Smith et al. 1996; McDowell and Schaefer 1996).

The first step in this analysis is to gain chemical-shift resolution in the solid state either by high-speed sample rotation about an axis tilted at the magic angle (which is half the tetrahedral angle) relative to the applied field, or by molecular orientation relative to the applied magnetic field. These resolution-enhancing techniques are often used in combination with stable-isotope labeling for additional chemical insight. The spinning experiments are typically performed on one micromole of (a) polycrystalline materials in contact with mother liquor, (b) frozen solutions, or (c) cryo- and lyoprotected lyophilized buffered glasses. With chemical-shift resolution in hand, structure and dynamics in the solid state are then inferred from relaxation measurements, just as in solution-state NMR, or determined by direct measurement of long-range, weak dipolar couplings in multidimensional experiments. The latter measurements are possible because the spatial averaging of weak dipolar coupling by magic-angle spinning can be inhibited by rotor-synchronized radiofrequency pulses in so-called recoupling experiments designed for either heteronuclear or homonuclear pairs of spins. The goal of the continuing development of these experiments is to offer solid-state NMR to the biological community as a practical structural alternative to X-ray crystallography and solution-state NMR.

2.2 Averaging of Dipolar Couplings

A 10-Å structural range is accessed by measuring the through-space dipolar coupling between two rare spins. The dipolar interaction between two spins depends on the inverse cube of the internuclear distance, on the orientation of the internuclear vector with respect to the applied static magnetic field, and on the magnetic coupling of the two nuclei (Stejskal and Memory 1994). That is, the dipolar coupling depends both on space and spin coordinates. Magic-angle sample spinning suppresses the dipolar interaction by averaging over the space coordinates. This averaging process can be defeated and the dipolar coupling partially restored by a competing averaging using rotor-synchronized radiofrequency pulses which operate exclusively on the spin coordinates. The space and spin averaging processes are both coherent and are of comparable frequency; therefore their combination produces destructive interference of averaging or recoupling. The extent of the interference is a measure of the dipolar coupling and hence the distance between nuclei.

2.3 Recoupling for Heteronuclear Spin Interactions

Rotational-echo double resonance (REDOR; Gullion and Schaefer 1989a) is a dephasing experiment for pairs of unlike nuclei that is performed in two parts (Fig. 1). In the first part, interfering pulses are present and a diminished signal (S) from one of the nuclei is detected. In the second part, the interfering pulses on the non-observed nucleus are omitted and a full signal (S_0) is detected. Events on the observed nuclear channel are identical in the two halves of the experiment. Thus, S/S_0 (or, equivalently, $\Delta S/S_0$, where $\Delta S = S_0 - S$) is a direct measure of the dipolar coupling between the unlike nuclei with no complications because of residual coupling to the protons. (Despite high-power decoupling, weak interactions with the protons remain.) REDOR measures ^{13}C-^{15}N (Gullion and Schaefer 1989b), ^{13}C-^{31}P (Hing et al. 1994; Beusen et al. 1995), ^{13}C-^{19}F (McDowell et al. 1996c), and ^{31}P-^{19}F (Studelska et al. 1996) distances up to 6, 8, 10, and 15 Å, respectively. Accuracies of a few tenths of an angstrom (McDowell et al. 1996e) to an angstrom are possible depending on sensitivity (Bennett et al. 1996) and the presence or absence of molecular motion (McDowell et al. 1996b).

REDOR (xy-8)

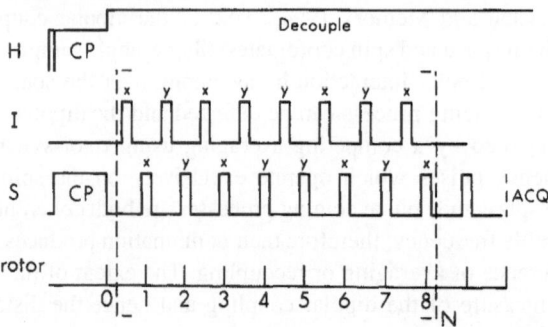

Fig. 1. REDOR pulse sequence with π pulses on both the I and S channels. The pulses are applied using an xy-phase-cycling scheme (Gullion et al. 1990) to suppress the effects of frequency offsets and compensate for pulse imperfections. Signal acquisition begins two rotor cycles after the completion of a full 8N rotor cycles of dephasing. The two extra rotor cycles (and a Hahn-echo refocusing pulse) are added to the sequence so that the start of data acquisition is not coincident with a pulse. A full-echo signal, S_o, is obtained by omitting the I-spin dephasing pulses. The dephased signal is S. The REDOR difference is $\Delta S = S_o - S$

2.4 Clusters of Labels

REDOR has been extended to combinations of several independent pairs in the same experiment so that multiple distances are determined (Holl et al. 1996). REDOR has also been extended to systems with one observed nucleus and multiple dephasing centers (McDowell et al. 1996b), and multiple observed nuclei coupled to one dephasing center (Wooley et al. 1997). In these applications, dipolar couplings between like nuclei must be much less than the magic-angle spinning speed. REDOR then defines an average coupling that is related to internuclear distances in terms of a model specifying relative orientations. For example, we calculated the expected REDOR dephasing of one S spin by a collection of two or more I spins using the powder average of a sum of independent dephasings: $S/S_o = <\Sigma_i \cos(\phi_{Di})_{\text{space and spin}} = \Pi_i \cos(\phi_{Di})>_{\text{space}}$, where ϕ_{Di} is the phase accumulation resulting from the heteronuclear dipolar coupling for the *ith* I-S pair; the first angle brackets indicate an average over space and spin coordinates, and the second angle brackets indicate just a spatial powder average. The average over spin coordinates includes dephasing by I-S spin pairs that are parallel (positive ϕ_D) and antiparallel (negative ϕ_D) and takes advantage of the additivity of arguments for the products of exponentials.

In the following sections, we will first describe an application of the REDOR methodology to a protein binding site whose structure is largely known from crystallography (EF-Tu). We will then compare predictions based on REDOR results with the crystal structure of a binding site of a system for which X-ray data become available after the NMR analysis had been completed (GlnBP). Next, we will illustrate the kind of detail that can presently be achieved using solid-state NMR on a liganded protein in combination with X-ray data on the *un*liganded protein (EPSP synthase). Finally, we will discuss the ability of REDOR to obtain useful binding-site information about complex, heterogeneous systems for which no crystallographic data are available (cell-wall components of *Staphylococcus aureus*).

2.5 EF-Tu

Elongation factor Tu (EF-Tu) is a 43-kDa bacterial protein belonging to the family of G proteins which act as molecular switches that are either on or off depending on whether the bound nucleotide is a diphosphate or triphosphate (Bourne et al. 1991). The role of EF-Tu in metabolism is to transport aminoacyl-tRNAs to the ribosome for protein biosynthesis. EF-Tu also serves as a subunit of the replicase of bacteriophage Qβ (Blumenthal et al. 1972) and as a site of action of several antibiotics including kirromycin, pulvomycin, and GE2270 (Anborgh and Parmeggiani 1991). EF-Tu is active when GTP is bound, and in that state complexes with all aminoacyl-tRNAs except fMet-tRNA^{fMet}. GTP hydrolysis at the ribosome initiates the release of EF-Tu from the ribosome. A second elongation factor binds to this complex resulting in the dissociation of GDP and subsequent binding of MgGTP by EF-Tu to complete the cycle.

The structure of EF-Tu has received considerable attention. The protein consists of three domains, with domain I containing the nucleotide binding site. A crystal structure of *T. thermophilus* EF-Tu com-

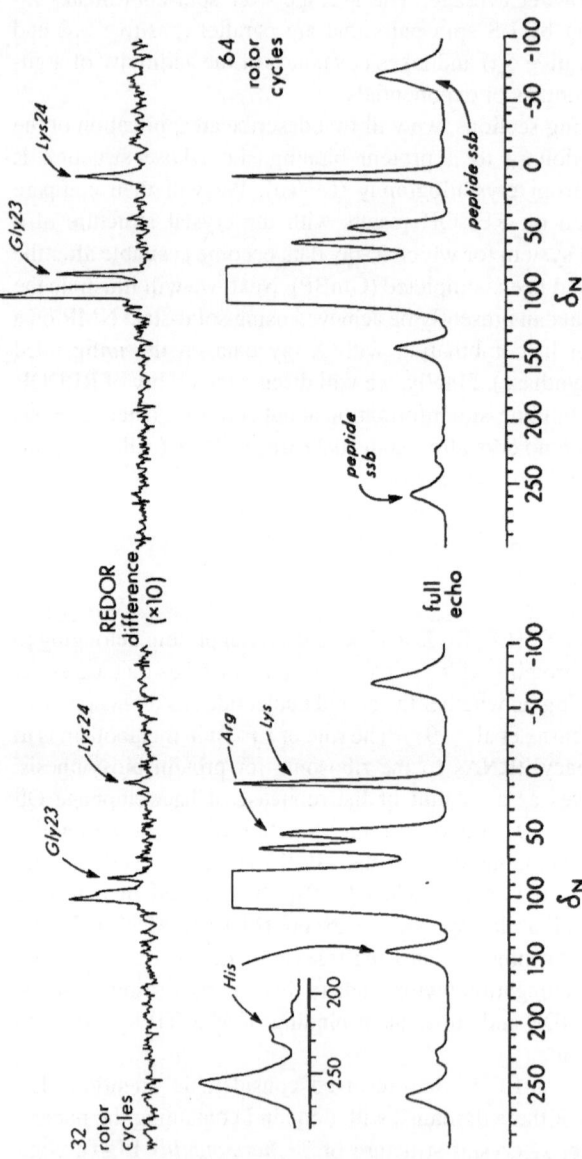

Fig. 2. REDOR ^{15}N NMR spectra of a lyophilized [*uniform*-^{15}N]-EF-TuMgGDP after 32 (*left*) and 64 (*right*) rotor cycles of ^{31}P dephasing with 5-kHz magic-angle spinning. The REDOR difference spectra (*top*) show peaks due to peptide nitrogens near ^{31}P (including a glycine peak near 80 ppm) and the ε-nitrogen peak of a lysine residue (9 ppm). Sidechain-nitrogen signals from histidine, arginine, and lysine residues are resolved in the S_0 full-echo spectra (*bottom*). Homogeneous decay of the peptide-nitrogen full-echo signal (and its spinning sidebands) is 11% faster (per 32 rotor cycles) than that of the lysine ε-nitrogen signal. Both S and S_0 spectra were the result of the accumulation of 27,648 scans

plexed with MgGDPMP has been obtained at a resolution of 1.7 Å
(Berchtold et al. 1993). Trypsinized *E. coli* EF-Tu with bound MgGDP,
missing a 15-amino acid loop from residues 43–59 in domain I, has been
solved to a resolution of 2.6 Å (Kjeldgaaard and Nyborg 1992). The
coordinates of domain I are available in the Brookhaven data base. The
nucleotide binding region consists of five parallel β strands and one
antiparallel β strand surrounded by α helices. A phosphate binding loop
from G18-K24 connects one of the β strands to an α helix.

We reported the results of REDOR NMR binding-site distance measurements made on the lyophilized EF-TuMgGDP complex isolated
from *E. coli* (McDowell et al. 1996a). Our objective in this work was to
compare distances obtained from solid-state NMR of the complete
EF-TuMgGDP complex with those obtained from X-ray crystallography of a complex that is missing a fragment in the binding region.
The REDOR difference spectra (Fig. 2, top) indicate that several peptide
nitrogens and at least one ε nitrogen of a lysine residue are near the ^{31}P
of GDP. Because glycine peptide nitrogens have a distinctive chemical
shift (Archer et al. 1993), the REDOR difference at 80 ppm must be due
to a glycine. From the X-ray structure (Kjeldgaard and Nyborg 1992),
we know that Gly23 is the only glycine in proximity of GDP. The only
lysine in proximity is Lys24. If no X-ray structure were available,
assigning a REDOR difference peak to a specific residue would be
difficult. Possible strategies include: (a) multidimensional NMR correlations; (b) site-specific chemical modification or mutagenesis; and (c)
site-specific stable-isotope labeling to produce unique dyads or triads of
labels. In the absence of a specific assignment, the identification of a
REDOR difference peak with a *single* residue is possible by observation
of the form of $\Delta S/S_0$ as a function of N_c (Mueller et al. 1995). We do not
view REDOR as a practical analytical method for generating total
structures for proteins and protein complexes. Instead, we believe that
the determination by REDOR of a few long-range distances imposes
hard constraints, which in *combination* with other structural information
(from either X-ray diffraction or solution-state NMR) can lead to unambiguous structural insights (Studelska et al. 1996).

The ΔS of Gly23 is twice that of Lys24 after 32 rotor cycles of
dephasing, and doubles in size after 64 rotor cycles, while the ΔS of
Lys24 triples in size (Fig. 2). Using the distances available from the
X-ray structure of EF-TuMgGDP missing residues 43–59 (Table 1), we

Table 1. REDOR dephasing for [*uniform*-^{15}N]EF-TuMgGDP

Residue	N_c	$\Delta S/S_0$ (observed)	$\Delta S/S_0$ (calculated with ^{31}P-^{31}P coupling)	$\Delta S/S_0$ (calculated without ^{31}P-^{31}P coupling)
Lys24	32	0.20[a]	0.205	0.209
	64	0.56[a]	0.624	0.638
Gly23	32	0.39[b]	0.496	0.508
	64	0.94[b]	1.02	1.04

[a]From Fig. 2, assuming 1 lysine contributes to ΔS, and 17 lysines contribute to S_0 (McDowell et al. 1996a).
[b]From Fig. 2, using 1/17 of the lysine full-echo peak height for S_0. This procedure avoids comparisons of the Gly23 REDOR difference and protein full-echo amide signals which differ considerably in shape and size (by three orders of magnitude). Homogeneous decay of the Gly23 S_0 is assumed to be that of the total peptide-amide peak (see caption to Fig. 2). If integrated intensities are used instead of peak heights, agreement between observed and calculated $\Delta S/S_0$ is reduced from within 10% to within 25%, although the relative dependence on N_c is unaffected.

can account for the observed REDOR dephasing to within 20%. Because of the inverse cube dependence of dipolar coupling and REDOR dephasing on distance, agreement of observed and calculated $\Delta S/S_0$ to within 20% means that the REDOR and X-ray determined distances are the same to within about 6%. For example, only Pβ contributes significantly to the dephasing of Lys24 (Table 1). Therefore, the experimental values cited in Table 1 for $\Delta S/S_0$ of 0.20 and 0.56 after 32 and 64 rotor cycles of dephasing, respectively, correspond to dipolar couplings of 70 and 64 Hz (Gullion and Schaefer 1989b). These couplings, which are not affected by ^{31}P-^{31}P interactions (Table 1), translate into ^{31}P-^{15}N distances of 4.05 and 4.17 Å, compared to the X-ray determined value of 4.03 Å. We conclude that in the vicinity of MgGDP, the structure of intact EF-TuMgGDP is indistinguishable from that of the complex that is missing residues 43–59, and that lyophilization has not altered the geometry of the binding site.

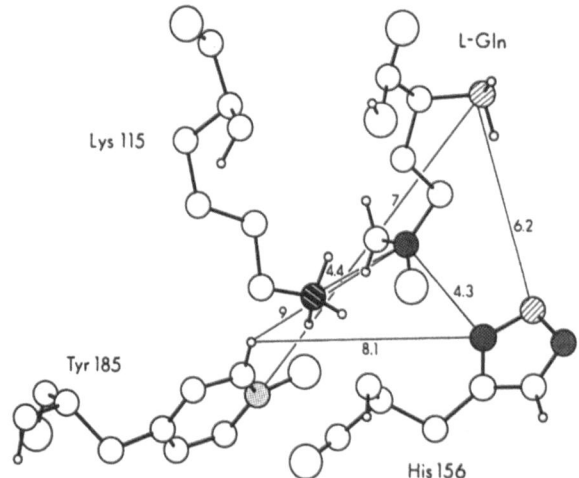

Fig. 3. Ball-and-stick representation of the binding site of glutamine binding protein showing REDOR-determined distances (in angstroms) between L-glutamine and residues Lys115, His156, and Tyr185. Labeled atoms (see Table 2) are *highlighted*

2.6 GlnBP

Glutamine binding protein (GlnBP) is a 25-kDa protein that is an essential component of the glutamine transport system in *E. coli.* We recently reported the determination of several internuclear distances in the complex between GlnBP and its ligand, L-glutamine (Hing et al. 1994), using stable-isotope labeling and REDOR. The REDOR-determined distances were also used as restraints in molecular dynamics calculations, which led to a proposed structure for the complex. This initial model was limited by the fact that the REDOR distances were measured from the ligand to residues in only one of the two GlnBP domains. In addition, the simulations were not sufficiently long to generate a stable closed structure. In subsequent work, we determined distances between L-glutamine and residues in both domains. We then compared these distances to those from an extended molecular dynamics simulation (Fig. 3), as well as to distances from the first report of an X-ray crystallograpic analysis of the complex (Hsiao et al. 1996).

Table 2. Comparison of observed and calculated internuclear distances for liganded glutamine binding protein

Dephasing (nucleus	Observing nucleus)	REDOR[a] (Å)	Model[b] (Å)	X-ray[c] (Å)
C_Δ of l-Gln	$N_{\epsilon 2}$ of His156	4.3 ± 0.2^d	4.4	3.7
$C_{\epsilon 1}$ of His156	N of l-Gln	6.2 ± 0.2^d	6.0	5.4
$C_{\epsilon 1}$ of His156	N_ϵ of l-Gln	7^d	6.9	5.2
C_Δ of l-Gln	N_ξ of Lys115	4.4 ± 0.2	5.6	3.5
C_Δ of l-Gln	$N_{\Delta 1}$ of His156	6.5	5.9	5.2
$F_{\epsilon 1}$ or $F_{\epsilon 2}$ of Tyr185	$N_{\epsilon 2}$ of His156	8.1 ± 0.5	8.2	8.7
$F_{\epsilon 1}$ or $F_{\epsilon 2}$ of Tyr185	$N_{\Delta 1}$ of His156	8.6 ± 0.5	9.5	8.6
$F_{\epsilon 1}$ or $F_{\epsilon 2}$ of Tyr185	C_Δ of l-Gln	9 ± 1^b	7.3	7.0
N of l-Gln	C_ξ of Tyr185	7 ± 1^b	6.6	4.4

[a]From Hing et al. (1994), unless otherwise noted.
[b]From Klug et al. (1997).
[c]Hsiao et al. (1996).
[d]REDOR constraint for simulation.

Even though the molecular dynamics simulations were restrained only by REDOR distances measured between L-glutamine and His156, the REDOR-measured distances to Lys115 (smaller domain) and Tyr185 (larger domain) are in reasonable agreement with all those obtained from the model, as well as with most of those reported by crystallographic analysis (Table 2). The most important exception is the distance from the amine nitrogen of the ligand and the oxygenated aromatic carbon of Tyr185. The 4.4-Å distance of the X-ray structure is hard to reconcile with the observed 2% ^{13}C-Tyr REDOR dephasing by amine-^{15}N labeled L-glutamine. At the moment, we have no explanation for this discrepancy.

Agreement between the REDOR-constrained molecular-dynamics model and the X-ray structure for hydrogen-bond formation is modest (Hsiao et al. 1996). In particular, Asp10, Arg75, and Asp157 appear to be misplaced by the modeling. The mechanism for closure of the GlnBP cleft seems to be the hydrogen-bond bridges that L-glutamine makes as the ligand forms an electrostatic shield between the two domains. These kinds of interactions are certain to be dependent on the details of the interaction potentials used in the simulation. Thus, without REDOR distances between L-glutamine and at least some of the residues forming hydrogen bonds as constraints, it is not surprising that the molecular dynamics simulations do not define the positions of Asp10, Arg75, and Asp157. On the other hand, the model and X-ray structure disagree on the placement of Tyr185, for which there is also a discrepancy between experimental REDOR and X-ray distances (Table 2). Perhaps future refinements of both REDOR-constrained simulations and X-ray structures will improve the overall agreement.

2.7 EPSP Synthase

2.7.1 Ternary Complex

The 46-kDa enzyme 5-enolpyruvylshikimate-3-phosphate (EPSP) synthase catalyzes the reversible condensation of shikimate-3-phosphate (S3P) and phosphoenolpyruvate (PEP) to form EPSP in the synthesis of aromatic amino acids in plants and microorganisms (Steinrucken and Amrhein 1984a; Anderson and Johnson 1990). This reaction is inhibited by the commercial herbicide N-(phosphonomethyl)glycine (glyphosate or Glp), $HO_3PCH_2NHCH_2COOH$, which, in the presence of S3P, binds to EPSP synthase (EPSPS) and forms a stable, ternary complex (Steinrucken and Amrhein 1984b; Anderson et al. 1988). A crystal structure of unliganded EPSP synthase has been published (Stallings et al. 1991), but there is no structure available for the ternary complex, which apparently does not form crystals suitable for diffraction studies. The crystal structure shows that EPSP synthase has two domains separated by a cleft which is presumably the region of the binding site.

2.7.2 Labeling Strategy

Our strategy for the NMR characterization of the ternary complex was in four parts. First, ^{13}C and ^{13}N labels were placed in Glp and distances were measured using REDOR NMR from these labels to ^{31}P in S3P and Glp (McDowell et al. 1996b). The measurements were performed on a ternary complex that had been frozen from dilute solution and lyophilized. Each EPSPS-S3P-Glp complex was surrounded by at least 50 buffer molecules so that the complex was immobilized in an ionic glass formed by the freeze-quenched buffer. Formation of the glass inhibits crystallization of ice within the binding site and prevents aggregation of the protein. Homogeneity of charge-stabilized binding sites of protein complexes embedded in ionic glasses has been observed in our laboratory for EPSP synthase (Christensen and Schaefer 1993), *E. coli* glutamine binding protein (Hing et al. 1994), and ribulose 1,5-bisphosphate carboxylase (Mueller et al. 1995). Distances resulting from inter- and intraligand EPSPS REDOR measurements helped to define the geometry of the two substrates relative to one another.

The second part of the strategy involved the introduction of ^{15}N labels into the lysine, arginine and histidine residues of EPSP synthase to determine which are near the ^{31}P of the negatively charged phosphate group of S3P and the carboxyl and phosphonate groups of Glp (McDowell et al. 1996e). The protein labeling was done using an *E. coli* expression system capable of producing 100 mg purified protein from 1 l defined media containing 100 mg/l ^{15}N-labeled amino acid. The third part of the strategy used mutagenesis to locate ^{19}F-labeled tryptophan residues in the upper and lower domains of EPSPS and long-range [19]F-^{31}P REDOR distance determinations to measure directly the extent to which the cleft closes around ^{31}P-containing ligands (Studelska et al. 1996). The fourth part of our strategy is in progress and will use all the interatomic distances determined by REDOR as restraints in molecular dynamics simulations to build a model of the ternary complex.

2.7.3 ^{15}N-Lysine Labeling

The phosphate group of S3P and the phosphonate group of Glp carry double negative charges, and the carboxyl carbon atoms of the two substrates each carry a single negative charge (Anderson et al. 1988; Stallings et al. 1991). Thus, of the order of six basic residues must be in the EPSPS binding site for charge balance. Some of these basic residues are likely to be lysine residues. There are 17 lysine residues in EPSP synthase, 4 of which are in the vicinity of the cleft according to the X-ray structure (Stallings et al. 1991). Figure 4 shows the CPMAS ^{15}N NMR Hahn-echo spectrum of [ε-^{15}N]Lys-EPSPS-S3P-Glp ternary complex (McDowell et al. 1996e). By comparison to the natural-abun-

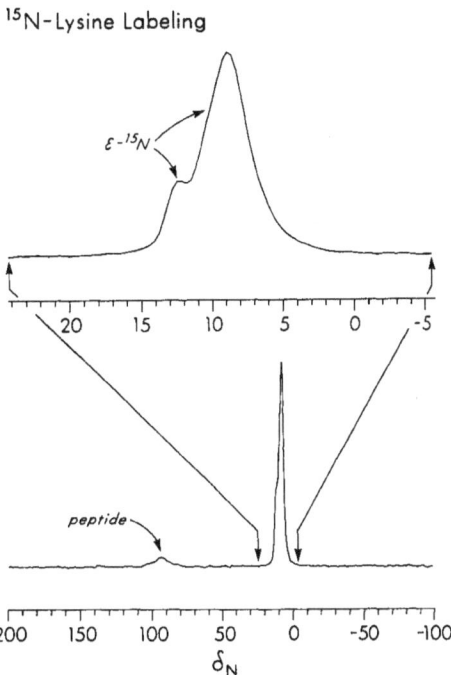

Fig. 4. CPMAS ^{15}N NMR Hahn-echo spectrum of a lyophilized [ε-^{15}N]Lys-EPSPS-S3P-Glp ternary complex. The EPSPS was overexpressed by genetically engineered *E. coli* grown on medium containing L-[ε-^{15}N]lysine. Isotopic incorporation in the ε nitrogen atoms of the lysine residues is 91% with no scrambling of label. The insert shows at least two chemical-shift environments for the lysine residues. The spectrum was the result of the accumulation of 8192 scans. Magic-angle spinning was at 5 kHz

dance amide-nitrogen peak (and taking into account the rotating-frame relaxation rates), we estimated that the ε nitrogen atoms of the lysine residues of EPSPS were 91(±5)% isotopically enriched (McDowell et al. 1996e). Proton-nitrogen cross-polarization transfer rates are consistent with the absence of large-amplitude motions for those lysine residue side-chain nitrogen atoms contributing to the observed signals of Fig. 4. A fraction of the lysine ^{15}N amine peak at 9 ppm is shifted to low field and partially resolved (Fig. 4, top). Shifted components are responsible

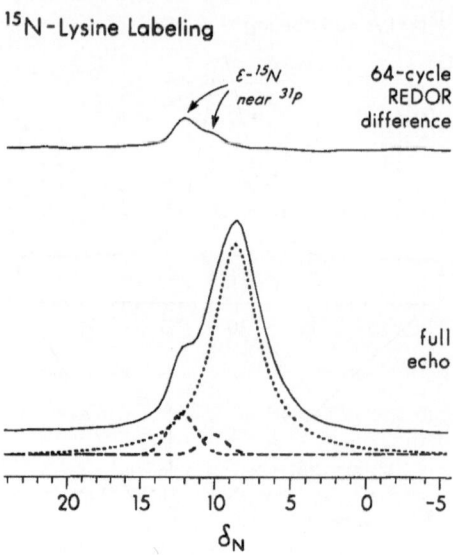

Fig. 5. REDOR ^{15}N NMR spectra of the ternary complex of **Fig. 4** after 64 rotor cycles of ^{31}P dephasing with 5 kHz magic-angle spinning. The REDOR difference spectrum (*top*) arises from at least two types of lysine residues whose resonances (11 and 13 ppm) are shifted from the main lysine peak at 9 ppm. Both S and So spectra were the result of the accumulation of 15,024 scans. *Broken lines* show the deconvolution of the full-echo spectrum assuming that three lysine residues contribute to the shifted peaks (see text)

for both of the lines observed in the REDOR difference spectrum (^{15}N observe, ^{31}P dephase; Fig. 5, top). An estimated spin count for these shifted lines suggests that three lysine residues are within 4 Å of the ^{31}P atoms of S3P and Glp. More involved NMR experiments confirmed this assignment and actually resolved the REDOR difference spectrum into three distinct peaks (McDowell et al. 1996e).

2.7.4 Residues in the Binding Site

Labeling with [^{15}N]arginine and [^{15}N]histidine yielded additional distances to S3P and Glp. The REDOR-determined distribution of labeled

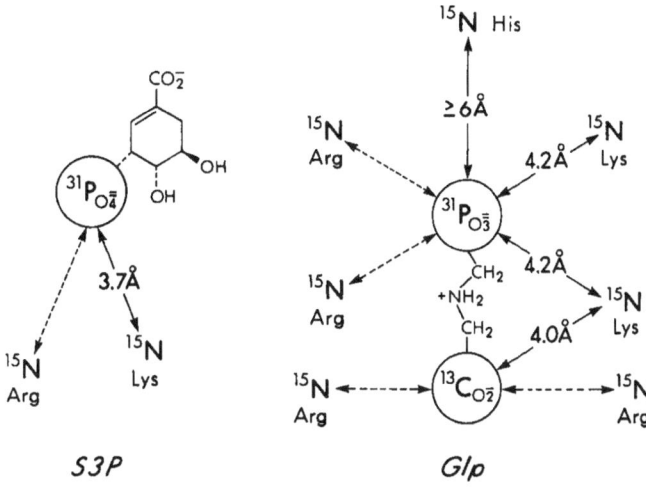

Fig. 6. Schematic representation of the orientation of S3P and Glp bound to the active site of EPSP synthase as determined by REDOR NMR. The phosphate group of S3P is charge stabilized by one arginine and one lysine residue. The phosphonate group of Glp is stabilized by two lysine residues, one of which is shared with the carboxyl group of Glp. Two arginine residues are near the phosphonate group of Glp and two additional arginine residues are near the carboxyl group. A single histidine residue is near the phosphonate group of Glp in the vicinity of the binding site but is not part of the charge balance

lysine, arginine and histidine residues in the binding site of the EPSP-synthase ternary complex is shown schematically in Fig. 6. From NMR data alone, we can only construct such a schematic diagram. While distances from the ^{31}P of Glp and S3P and the C-1 carbon of Glp to the ε-^{15}N lysine labels are accurately known from REDOR, the determination of ^{31}P and ^{13}C distances to the guanidino-^{15}N labels of Arg and the ring-^{15}N labels of His require modeling of side-group geometry to take into account the two ^{15}N dephasing centers in each residue side-chain (unpublished results). Nevertheless, the observed extensive dephasings (McDowell et al. 1996e) mean that the distances from Glp to the arginine ^{15}N labels are short, of the order of 4 Å. Thus, the two arginine residues that are proximate to the phosphate group cannot be the same two that are near the carboxyl group. Altogether we have identified seven basic residues participating in the ternary-complex binding-site charge balance: three lysine and four arginines. There may be others around the carboxyl carbon atom of S3P. The single histidine residue near glyphosate is more distant than any of the surrounding lysine and arginine residues, while the arginine residue in the vicinity of the S3P phosphate group may actually be one of the four arginine residues clustered about glyphosate.

2.7.5 ^{19}F-Tryptophan Labeling

We used site-directed mutagenesis to remove and introduce tryptophans into an *E. coli* isoform of EPSP synthase (Studelska et al. 1996). Clones were screened by restriction analysis and the absence of errors in the EPSP synthase coding region of each final plasmid construct was confirmed by commercial DNA sequencing. Fluorine label was biosynthetically introduced into the enzyme via DL-[6-^{19}F]tryptophan using bacteria auxotrophic for tryptophan. The positioning of the ^{19}F labels is illustrated using the structure shown in Fig. 7 (left). Wild-type EPSP synthase has only two tryptophan residues (W289 and W337) and both are in the lower domain (Stallings et al. 1991). We created a single mutant (W289Q) by removing the lower-domain tryptophan, which was the more distant of the two from the cleft region. We then created a double mutant (W289Q;F172 W) by introducing tryptophan into the upper domain near the binding site. When the tryptophans were specifically labeled by ^{19}F, measurements of long-range ^{19}F-^{31}P monitored directly the extent of closure of the cleft on binding of ^{31}P-containing ligands (Fig. 7, right). The separation between tryptophans in the upper and lower domain decreases by about 20 Å. We believe that the closed structure is the functional form in solution, and that the open jaws of the unliganded crystal structure are an artifact resulting from the conditions used to create a suitable crystal for diffraction.

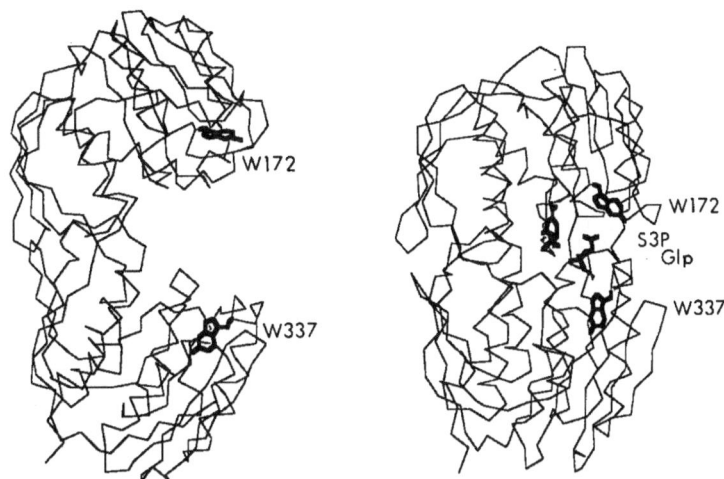

Fig. 7. *Left* Cα trace of 5-enolpyruvylshikimate-3-phospate synthase, based on the X-ray diffraction (Stallings et al. 1991), showing the approximate locations of W172 (*upper domain*) and W337 (*lower domain*) of the double mutant W289Q;F172 W. The side-chain of F172 in the wild-type structure can be modified to tryptophan with retention of the wild-type rotameric state and only slight adjustment to remove van der Waals contacts with nearby residues. The F-F distance in this structure is 38 Å. *Right* Cα trace of a model of the EPSPS ternary complex consistent with ^{19}F-^{31}P REDOR distance measurements (Studelska et al. 1996). Ligands were docked into the X-ray structure based on the results of REDOR studies involving complexes of specifically ^{13}C- and ^{15}N-labeled protein and ligands (McDowell et al. 1996b,d). The hinge was identified by searching for residues in the connecting strands that were not part of a secondary structural element in either the upper or lower domain. Torsions in the two hinge strands in EPSPS were manually adjusted to bring the two protein domains into proximity. Ligand position and hinge torsions were adjusted iteratively to accommodate P-F, C-N, C-P, and N-P distances. The F-F distance in this model is 21 Å. The F-P distances for W337-S3P, W337-Glp, W172-S3P, and W172-Glp in the model are 18, 13, 7, and 11 Å, respectively

S.*Aureus* PEPTIDOGLYCAN

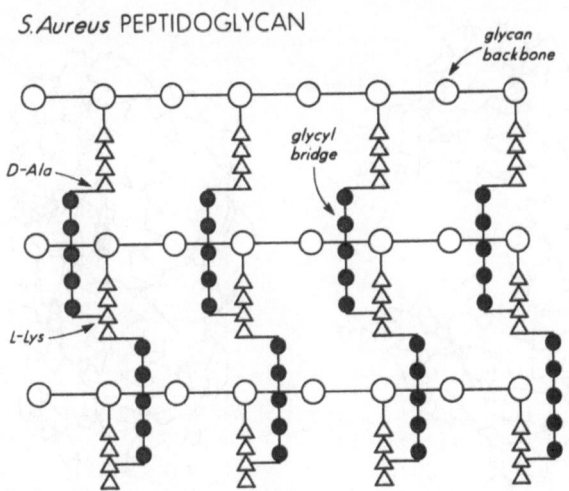

Fig. 8. Schematic representation of an idealized version of the cell-wall peptidoglycan of *S. aureus* (after Stryer). A four-unit peptide stem (*triangles*) having the sequence, l-Ala-d-Glu-l-Lys-d-Ala, is attached to every second sugar
of the glycan backbone (*open circles*). Cross-linking between glycans occurs
through pentaglycyl bridges (*dark circles*) connecting the carbonyl carbon of
d-Ala of the fourth position of one stem and the εnitrogen of l-Lys of the third
position of another

2.8 *Staphylococcus aureus*

2.8.1 Bacterial Cell-Wall Peptidoglycan

The major component of the bacterial cell wall is peptidoglycan, a
cage-like macromolecule which covers the entire cell (Rogers et al.
1980). The general chemical structure of peptidoglycan is the same for
all bacterial species and consists of a glycan backbone of polysaccharide
chains with a repeat unit of *N*-acetylglucosamine and *N*-acetylmuramic
acid linked by a β (1–4) covalent bond (Figs. 8, 9). Peptide stems are
attached to the *N*-acetylmuramic acid units of the backbone, and these
stems differ from one species to another. The peptidoglycan of *Staphylococcus aureus* (A3α) is based on the pentapeptide stem L-Ala-D-Glu-
L-Lys-D-Ala-D-Ala. Cross-linking occurs by means of pentaglycyl

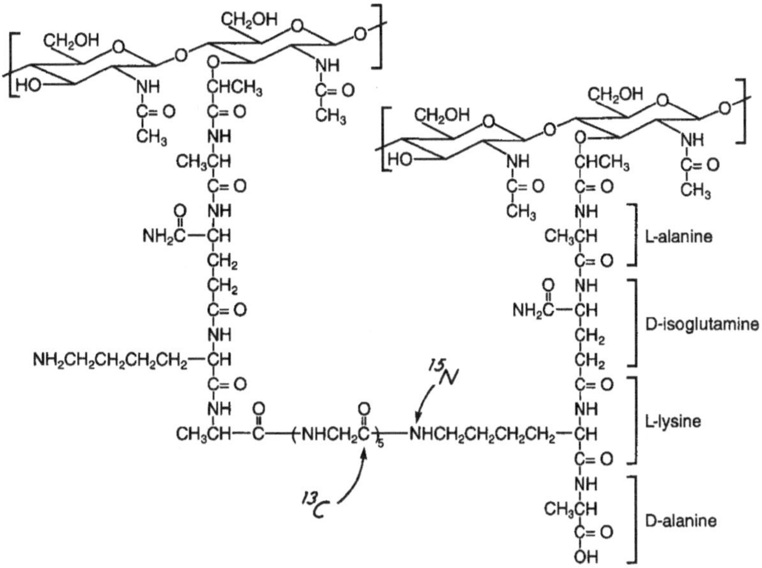

Fig. 9. Chemical structure of the peptidoglycan of *S. aureus*, with two sites identified which are suitable for stable-isotope labeling

bridges from the carbonyl carbon of D-Ala of the fourth position of one stem to the ε nitrogen of L-Lys of the third position of another. Terminal D-Ala units of cross-linked stems are eliminated (Rogers et al. 1980). Cross-linking presumably adds to the structural integrity of the bacterial cell wall and may also affect its permeability (Ghuysen et al. 1968). Because the cell changes volume as much as 200% with changing ionic conditions (Marquis 1968; Ou and Marquis 1970), the cell wall must be flexible. One model of the cell wall (Leps et al. 1987) proposes that the glycan chains confer structural rigidity and that the cross-linking bridges provide flexibility by expanding under stress (see Fig. 8). We used solid-state NMR to define the structure and dynamics of bacterial cell-wall peptidoglycan in situ. This work is part of a program whose goal is to aid in the design and discovery of new antibiotics that interfere with the synthesis or integrity of the bacterial cell wall.

2.8.2 REDOR of the Pentaglycyl Bridge

Characterization of the conformation of the pentaglycyl bridges is possible by REDOR NMR of cell walls labeled by ^{13}C and ^{15}N. We either observe a single ^{15}N under the influence of coupling to five ^{13}C's (Fig. 10, left), or we observe five independent ^{13}C's, each coupled to the same ^{15}N (Fig. 10, right). In both situations, the REDOR dephasing depends on the geometry of the bridge. For the latter analysis with labeling by [1-^{13}C]glycine and L-[6-^{15}N]lysine, we assume that the ^{13}C full-echo REDOR signal is due to five ^{13}C-labeled glycyl carbonyl carbons, one of which is directly bonded to the 100% ^{15}N-enriched ε

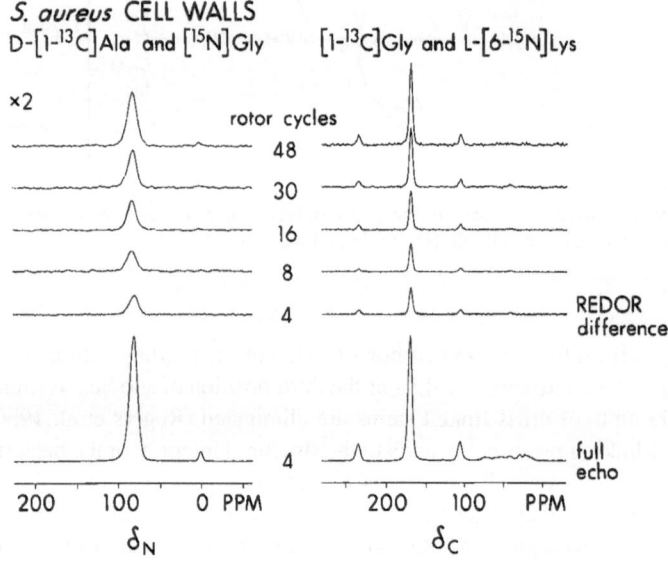

Fig. 10. *Left* REDOR ^{15}N NMR spectra (with ^{13}C dephasing) of cell walls of *S. aureus* labeled by D-[1-^{13}C]alanine and [^{15}N]glycine. The REDOR differences are shown as a function of the number of rotor cycles of dephasing. The full-echo spectrum after four rotor cycles with no dephasing is *at the bottom of the figure*. The REDOR differences have been scaled so that $\Delta S/S_0$ is determined by comparison to the full-echo signal for $N_c = 4$. *Right* REDOR ^{13}C NMR spectra (with ^{15}N dephasing) of cell walls labeled by [1-^{13}C]glycine and L-[6-^{15}N]lysine. Both left and right sets of spectra were obtained with 3.205-kHz magic-angle spinning

nitrogen of the cross-linking lysyl side-chain (Tong et al 1997; Goetz and Schaefer 1997). Thus, the directly bonded ^{13}C-^{15}N pair accounts for the $^{13}C\{^{15}N\}$ REDOR difference of 0.2 (relative to the full-echo intensity) after four rotor cycles of dephasing with 3.205-kHz magic-angle spinning (Fig. 10, right). The REDOR difference grows to about 0.6 after only 48 rotor cycles, which indicates that most of the glycyl carbonyl labels are within 5 Å (Gullion and Schaefer 1989b; Marshall et al. 1990) of the cross-linked lysyl ^{15}N-labeled side-chain. That is, the glycyl bridges must be in a compact conformation.

2.8.3 Whole Cells

The dependence of the $^{13}C\{^{15}N\}$ REDOR dephasing ($\Delta S/S_o$) on the total evolution time (Gullion and Schaefer 1989b) is shown in Fig. 11 for lyophilized cell walls, and for lyophilized and hydrated whole cells, labeled by [1-^{13}C]glycine and L-[6-^{15}N]lysine (Fig. 9). Because the ^{13}C-^{15}N pair detected by REDOR is for an isopeptide bond occurring only in cross-links of peptidoglycan (the ϵ-^{15}N lysyl label does not scramble), whole-cell REDOR spectra have the same selectivity as cell-wall REDOR spectra (Tong et al. 1997). The $\Delta S/S_0$ of the hydrated whole cells at $-10°C$, and of the lyophilized whole cells and cell walls, are those of a compact conformation (Labischinski et al. 1993) with the five carbonyl-carbon to ϵ-nitrogen ^{13}C-^{15}N distances comparable to those of an α helix. The observed $\Delta S/S_0$ would be equally well matched by couplings calculated for a 3_1 helix, which is more likely for glycine-rich peptides (Saitô 1986). An extended β-strand conformation, with carbon-nitrogen distances of up to 15 Å, can only account for some 30% total dephasing (Hirsh et al. 1996), whereas almost 80% dephasing is observed. Thus, while we are unable to determine exactly what kind of compact conformation the bridge is in, we can conclude that the average bridge conformation is not extended.

The fact that five ^{13}C-^{15}N couplings are needed to fit the $\Delta S/S_0$ dephasing is consistent with complete glycyl bridges. The hydrated whole cells at $0°C$ (solid diamonds) have about half the $\Delta S/S_0$ dephasing of the frozen whole cells (Fig. 11). We attribute the reduction in dephasing to an increase in bridge motion (Tong et al. 1997). Regardless of the details of the motional averaging, the fact remains that despite

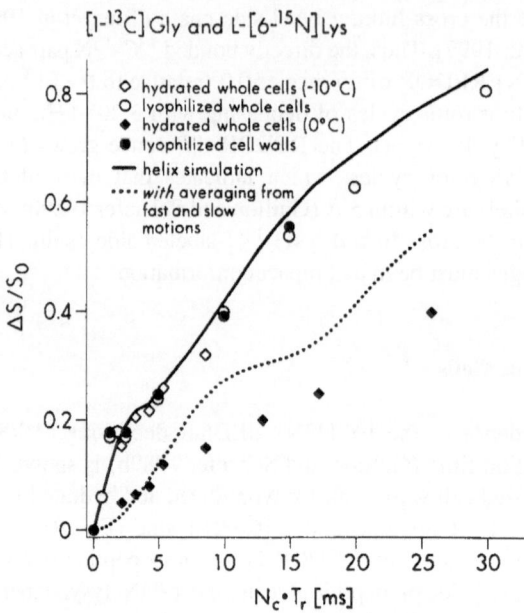

Fig. 11. REDOR dephasing ($\Delta S/S_0$) as a function of the evolution time ($N_C T_r$) for whole cells and cell walls of *S. aureus* labeled by [1-^{13}C]glycine and 1-[6-^{15}N]lysine (Tong et al. 1997). Results for lyophilized samples are *represented by circles*, and for hydrated samples *by diamonds*

extensive motion for the hydrated whole cells at 0°C, almost half of the REDOR dephasing that is associated with lyophilized cell walls is still observed for the hydrated whole cells. Thus, the average conformation must be compact. We conclude that although extension of the cell-wall peptidoglycan may occur under stress, the time-averaged equilibrium peptidoglycan conformation for at least 90% of the hydrated cell walls is compact, not extended. The remaining 10% may be in a more flexible, extended conformation and the two populations could be in slow exchange. The appearance of an upfield β-strand shift in whole-cell spectra at 0°C (Tong et al. 1997) supported the presence of a small fraction of extended conformations.

2.8.4 Vancomycin-Resistant Staphylococci

Vancomycin and other closely related glycopeptide antibiotics are used in the treatment of serious infection due to gram-positive bacteria (Reynolds 1989). The importance of vancomycin has been increasing in the last few years due to the emergence of methicillin-resistant staphylococci, including strains of *S. aureus* (Sieradzki and Tomasz 1996). Vancomycin exerts antibacterial action by the inhibition of peptidoglycan biosynthesis (Reynolds and Somner 1990). The antibiotic forms a complex with the D-Ala-D-Ala terminus of the peptidoglycan precursor, a lipid-linked *N*-acetylglucosamine-*N*-muramyl-pentapeptide. Recently, vancomycin-resistant enterococci have emerged (Reynolds and Snaith 1994). Resistance mediated by the *van* gene cluster is due to two mutant enzymes which result in a precursor peptidoglycan pentapeptide with a D-Ala-D-Lac terminus. This alteration reduces the binding affinity of vancomycin (Liu et al. 1994).

2.8.5 Working Model of Glycopeptide Mode of Action

Solution-state NMR, largely performed by Dudley Williams and his coworkers at the University of Cambridge (Williams 1996; Wright and Walsh 1992), has been the main structural tool supporting biochemical evidence that vancomycin does not directly interfere with a cell-wall enzyme, but rather complexes a peptidoglycan precursor thereby stopping cell-wall synthesis. The NMR studies were of vancomycin complexes with model peptides and showed that binding of vancomycin to D-Ala-D-Ala involves the formation of five hydrogen bonds. This hydrogen-bond network is perturbed by the replacement of the pentapetide terminal D-alanine by D-lactate, which reduces vancomycin binding affinity. Although these vancomycin derivatives have only a modestly increased affinity for model peptides in solution, they apparently have a greatly increased affinity for precursor binding in situ, as inferred from enhanced potency and uptake. This situation emphasizes the appropriateness of examining vancomycin complexes directly with solid-state NMR. Williams has suggested that the tight binding in situ is due to the "Gulliver Effect" (Williams 1996; Beauregard et al. 1995). Binding is enhanced by a combination of intermolecular interactions between the antibiotic (dimer formation) and between the antibiotic and the cell wall (membrane anchoring), leading to an overall reduced molecular mobility. The vancomycin derivative is immobilized just as Gulliver was by the Lilliputians: by a *combination* of many, weak tie lines.

2.8.6 Solid-State NMR Detection of Cell-Wall Complexes

Attractive as the Gulliver explanation is, it is just a cartoon suggestion of what may be happening within the cell wall. We propose to test the model with experimental results from solid-state NMR. The sensitivity of REDOR experiments on whole-cell and cell-wall samples of *S. aureus* will be exploited in new work in which we will detect complexes of the antibiotic vancomycin with cell-wall precursors in situ. These experiments will be designed to elucidate the mode of action of vancomycin. We will use a vancomycin-producing microorganism (*Streptomyces orientalis* C329) that grows on defined media and produces 1 g vancomycin/l medium. Thus, we have the ability to ^{13}C and ^{15}N-label vancomycin itself, just as we have labeled *S. aureus*. With this labeling capability, we propose to examine five aspects of vancomycin-*S. aureus* cell-wall complex formation: (a) the vancomycin-pentapeptide binding site; (b) the vancomycin cell-wall attachment; (c) the vancomycin intermolecular dimer interface; (d) positioning of the vancomycin-derivative membrane anchor; and (e) molecular mobilities of vancomycin and ^{19}F-labeled vancomycin-derivative cell-wall complexes.

This work in progress on *S. aureus* is illustrative of our approach to solid-state NMR of biological materials. Our goal is not the determination of total structure, but rather the analysis of restricted regions (interfaces, channels, binding sites) important to biological function. This approach uses: (a) REDOR methods for internuclear distances as great as 15 Å; (b) specific stable-isotope labeling schemes for three and four different kinds of nuclei; and (c) NMR probes tuned simultaneously to as many as six radiofrequencies for the detection of clusters of stable-isotope labels.

Acknowledgements. This work was supported by NIH grant GM40634. Permission is gratefully acknowledged to reproduce Figs. 1, 4–6 (Academic Press), Fig. 2 (Elsevier Press), and Figs. 3, 7, and 8–11 (American Chemical Society).

References

Anborgh PH, Parmeggiani A (1991) New antibiotic that acts specifically on the GTP-bound form of elongation factor Tu. EMBO J 10:779–784

Anderson KS, Johnson KA (1990) Kinetic and structural analysis of enzyme intermediates: lessons from EPSP synthase. Chem Rev 90:1131–1149

Anderson KS, Sikorski JA, Johnson KA (1988) Evaluation of 5-enolpyruvoylshikimate-3-phosphate synthase substrate and inhibitor binding by stopped-flow and equilibrium fluorescence measurements. Biochemistry 27:1604–1610

Archer SJ, Bax A, Roberts AB, Sporn MB, Ogawa Y, Piez KA, Weatherbee JA, Tsang ML-S, Lucas R, Zheng B-L, Wenker J, Torchia DA (1993) Transforming growth factor β1: NMR signal assignments of the recombinant protein expressed and isotopically enriched using Chinese hamster ovary cells. Biochemistry 32:1152–1163

Beauregard DA, Williams DH, Gwynn MN, Knowles DJC (1995) Dimerization and membrane anchors in extracellular targeting of vancomycin group antibiotics. Antimicrob Agents Chemother 39:781–785

Bennett AE, Rienstra CM, Auger M, Lakshmi KV, Griffin RG (1995) Heteronuclear decoupling in rotating solids. J Chem Phys 103:6951–6958

Berchtold H, Reshetnikova L, Reiser COA, Schirmer NK, Sprinzl M, Hilgenfeld R (1993) Crystal structure of active elongation factor Tu reveals major domain rearrangements. Nature 365:126–132

Beusen DD, McDowell, LM, Slomczynska U, Schaefer J (1995) Solid-state NMR analysis of the conformation of an inhibitor bound to thermolysin. J Med Chem 38:2742–2747

Blumenthal T, Landers TA, Weber K (1972) Bacteriophage Qβ replicase contains the protein biosynthesis elongation factors EF/Tu and EF/Ts. Proc Natl Acad Sci USA 69:1313–1317

Bourne HR, Sanders DA, McCormick F (1991) The GTPase superfamily: conserved structure and molecular mechanism. Nature 349:117–127

Christensen AM, Schaefer J (1993) Solid-state NMR determination of intra- and intermolecular ^{31}P-^{13}C distances for shikimate 3-phosphate and [1-^{13}C]glyphosate bound to enolpyruvylshikimate-3-phosphate synthase. Biochemistry 32:2868–2873

Ghuysen J-M, Strominger JL, Tipper DJ (1968) Bacterial cell walls. In: Florkin M, Stotz EH (eds) Comprehensive biochemistry, vol 26A. Elsevier, Amsterdam, pp 53–104

Goetz JM, Schaefer J (1997) REDOR dephasing by multiple spins in the presence of molecular motion. J Magn Reson 127:147–154

Griffiths JM, Griffin RG (1993) Nuclear magnetic resonance methods for measuring dipolar couplings in rotating solids. Anal Chim Acta 283:1081–1101

Gullion T, Schaefer J (1989a) Rotational-echo double-resonance NMR. J Magn Reson 81:196–200

Gullion T, Schaefer J (1989b) Detection of weak heteronuclear dipolar coupling by rotational-echo double-resonance NMR. Adv Magn Reson 13:57–83

Gullion T, Baker DB, Conradi MS (1990) New, compensated Carr-Purcell sequences. J Magn Reson 89:479–484

Hing AW, Tjandra N, Cottam PF, Schaefer J, Ho C (1994) An investigation of the ligand-binding site of the glutamine-binding protein of *Escherichia coli* using rotational-echo double-resonance NMR. Biochemistry 33:8651–8661

Hirsh DJ, Hammer J, Maloy WL, Blazyk J, Schaefer J (1996) Secondary structure and location of a magainin analogue in synthetic phospholipid bilayers. Biochemistry 35:12733–12741

Holl SM, Kowalewski T, Schaefer J (1996) Characterization of two forms of cadmium phosphide by magic-angle-spinning P-31 NMR. Solid State NMR 6:39–46

Hsiao C-D, Sun Y-J, Rose J, Wang B-C (1996) The crystal structure of glutamine-binding protein from Escherichia coli. J Mol Biol 262:225–242

Kjeldgaard M, Nyborg J (1992) Refined structure of elongation factor EF-Tu from *Escherichia coli*. J Mol Biol 223:721–742

Klug CA, Tasaki K, Tjandra N, Ho C, Schaefer J (1997) Closed form of liganded glutamine-binding protein by rotational-echo double-resonance NMR. Biochemistry 36:9405–9408

Labischinski H, Hochberg M, Sidow T, Maidhof H, Henze U, Berger-Bächi B, Weche J (1993) Biophysical and biochemical studies on the fine structure of the sacculi from *Escherichia coli* and *Staphylococcus aureus*. In: DePedro MA, Höltje JV, Löffelhardt W (eds) Bacterial growth and lysis: metabolism and structure of the bacterial sacculus. Plenum, New York, pp 9–21

Leps B, Labischinski H, Bradaczek H, (1987) Conformational behavior of the polysaccharide backbone of murein. Biopolymers 26:1391–1406

Liu J, Volk KJ, Lee MS, Pucci M, Handwerger S (1994) Binding studies of vancomycin to the cytoplasmic peptidoglycan precursors by affinity capillary electrophoresis. Anal Chem 66:2412–2416

Marquis RE (1968) Salt-induced contraction of bacterial cell walls. J Bacteriol 95:775–781

Marshall GR, Beusen DD, Kociolek K, Redlinski AS, Leplawy MT, Pan Y, Schaefer J (1990) Determination of a precise interatomic distance in a helical peptide by REDOR NMR. J Am Chem Soc 112:963–966

McDowell LM, Schaefer J (1996) High-resolution NMR of biological solids. Curr Opin Struct Biol 6:624–629

McDowell LM, Barkan D, Wilson GE, Schaefer J (1996a) Structural constraints on the complex of elongation-factor Tu with magnesium guanosine diphosphate from rotational-echo double-resonance NMR. Solid State Nucl Magn Reson 7:203–210

McDowell LM, Klug CA, Beusen DD, Schaefer J (1996b) Ligand geometry of the ternary complex of 5-enolpyruvylshikimate-3-phosphate synthase from rotational-echo double-resonance NMR. Biochemistry 35:5395–5403

McDowell LM, Lee M, McKay RA, Anderson KS, Schaefer J (1996c) Intersubunit communication in tryptophan synthase by carbon-13 and fluorine-19 REDOR NMR. Biochemistry 35:3328–3334

McDowell LM, Schmidt A, Cohen ER, Studelska DR, Schaefer J (1996d) Structural constraints on the ternary complex of 5-enolpyruvylshikimate-3-phosphate synthase from rotational-echo double-resonance NMR. J Mol Biol 256:160–171

Mueller DD, Schmidt A, Pappan KL, McKay RA, Schaefer J (1995) Activator carbamino carbon to inhibitor phosphorus internuclear distances in ribulose-1,5-bisphosphate carboxylase/oxygenase. A solid state NMR study. Biochemistry 34:5597–5603

Ou L-T, Marquis RE (1970) Electromechanical interactions in cell walls of gram-positive cocci. J Bacteriol 101:92–101

Reynolds PE (1989) Structure, biochemistry and mechanism of action of glycopeptide antibiotics. Eur J Clin Microbiol Infect Dis 8:943–950

Reynolds PE, Snaith HA, Maguire AJ, Dutka-Malen S, Courvalin P (1994) Analysis of peptidoglycan precursors in vancomycin-resistant *Enterococcus gallinarum* BM4174. Biochem J 301:5–8

Reynolds PE, Somner EA (1990) Comparison of the target sites and mechanisms of action of glycopeptide and lipoglycodepsipetide antibiotics. Drugs Exp Clin Res 16:385–389

Rogers HJ, Perkins HR, Ward JB (1980) Microbial cell walls and membranes. Chapman and Hall, London

Saitô H (1986) Conformation-dependent ^{13}C chemical shifts: a new means of conformational characterization as obtained by high-resolution solid-state ^{13}C NMR. Magn Reson Chem 24:835–852

Sieradzki K, Tomasz A (1996) A highly vancomycin-resistant laboratory mutant of Staphylococcus aureus. FEMS Microbiol Lett 142:161–166

Smith SO, Aschheim K, Groesbeek M (1996) Magic-angle-spinning NMR-spectroscopy of membrane-proteins. Q Rev Biophys 29:395–449

Stallings WC, Abdel-Meguid SS, Lim LW, Shieh H-S, Dayringer HE, Leimgruber NK, Stegeman RA, Anderson KS, Sikorski JA, Padgette SR, Kishore GM (1991) Structure and topological symmetry of the glyphosate target 5-enolpyruvylshikimate-3-phosphate synthase: a distinctive protein fold. Proc Natl Acad Sci USA 88:5046–5050

Steinrücken HC, Amrhein N (1984a) 5-Enolpyruvylshikimate-3-phosphate synthase of *Klebsiella pneumoniae*. 1. Purification and properties. Eur J Biochem 143:341–349

Steinrücken HC, Amrhein N (1984b) 5-Enolpyruvylshikimate-phosphate synthase of *Klebsiella pneumoniae*. 2. Inhibition by glyphosate [N-(phosphonomethyl)glycine]. Eur J Biochem 143:351–357

Stejskal EO, Memory JD (1994) High resolution NMR in solid state: fundamentals of CP/MAS. Oxford University Press, New York, p 65

Studelska DR, Klug CA, Beusen DD, McDowell LM, Schaefer J (1996) Long-range distance measurements of protein binding sites by REDOR NMR. J Am Chem Soc 118:5476–5477

Tong G, Pan Y, Dong H, Pryor R, Wilson GE, Schaefer J (1997) Structure and dynamics of pentaglycyl bridges in the cell walls of *Staphylococcus aureus* by ^{13}C-^{15}N REDOR NMR. Biochemistry 36:9859–9866

Williams DH (1996) The glycopeptide story – how to kill the deadly "super-bugs." Nat Prod Rep 469–477

Wooley KL, Klug CA, Tasaki K, Schaefer J (1997) Shapes of dendrimers from rotational-echo double-resonance NMR. J Am Chem Soc 119:53–58

Wright GD, Walsh CT (1992) D-alanyl-D-alanine ligases and the molecular mechanism of vancomycin resistance. Acc Chem Res 25:468–473

3 Molecular and Dendritic Receptors for Small Biomolecules

F. Diederich

3.1   Introduction 53
3.2   Cage- and Cleft-Type Receptors for *N*-Protected Excitatory
        Amino Acids 54
3.3   Carbohydrate Recognition 63
3.4   Double-Decker Cyclophane Receptors for Steroids:
        Dissolution of Cholesterol in Water 66
3.5   Dendrophanes: Dendritic Receptors for the Complexation
        of Arenes and Steroids in Aqueous Solution 72
3.6   Conclusions 76
References 77

3.1 Introduction

Molecular recognition phenomena are at the heart of highly selective chemical reactions and transport phenomena, as well as the formation of functional supramolecular assemblies in living systems. Studies of synthetic model systems complement biological investigations in contributing to the understanding of these processes and, at the same time, offer new mechanisms for controlling reactivity and specificity in chemistry (Lehn 1995). These fascinating perspectives have guided our research in molecular recognition with synthetic receptors over the past 17 years (Diederich 1991). The detailed insight into the nature and strength of weak intermolecular bonding interactions as well as molecular shape complementarity gained from these studies provided a strong fundament for our recently initiated program targeting the structure-based de novo design of nonpeptidic antagonists for therapeutically relevant protein receptors (Obst et al. 1997).

In this article, we describe our development of selective receptors for excitatory amino acid derivatives, carbohydrates, and steroids. We selected excitatory amino acid derivatives as target substrates since X-ray crystal structures of neuroreceptors recognizing excitatory amino acids have not yet been solved (Krogsgaard-Larsen and Hansen 1992; Watkins and Collingridge 1989) and, therefore, the principles governing their selective complexation in biology remain to be explored. Although many X-ray crystal structures of carbohydrate binding proteins bound to their substrates have been solved since 1984, the extraordinary complexity and multiplicity of H-bonding and *van-der-Waals* contacts in these complexes has largely prevented a detailed analysis of individual bonding interactions (Lemieux 1991; Quiocho 1989; Sharon and Lis 1993; Wong et al. 1995). The first X-ray structural investigations on biological steroid complexes started appearing around 1990 (Wallimann et al. 1997) and displayed large hydrophobic binding sites often surrounded by side chains of aromatic amino acids (Arevalo et al. 1993). Since these binding sites resemble the cavities of large cyclophane receptors (Diederich 1991; Murakami et al. 1996; Vögtle 1990), we felt that the development of highly specific, tight-binding steroid receptors based on cyclophane scaffolds could potentially lead to a new class of pharmacological agents displaying functions such as steroid solubilization, enhanced steroid transport and delivery, and steroidal drug stabilization, which had previously been exclusively targeted with cyclodextrins (Szejtli 1988; Wallimann et al. 1997). In the last section of the article, we show that the introduction of cyclophane receptors as initiator cores into globular dendrimers (Newkome et al. 1996) generates efficient model systems for biological recognition sites that are deeply buried within globular proteins.

3.2 Cage- and Cleft-Type Receptors for *N*-Protected Excitatory Amino Acids

The cage-like chiral C_3-symmetrical receptor (S,S,S)-(+)-**1** (Fig. 1) was prepared by a short, convergent synthesis (Pieters et al. 1997). It con-

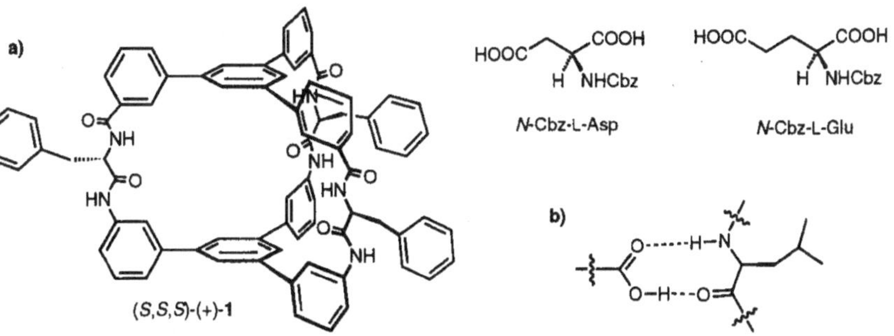

Fig. 1. a Cage-like, C_3-symmetrical receptor for the enantioselective recognition of the *N*-Cbz-protected excitatory amino acids aspartic acid (Asp) and glutamic acid (Glu), and **b** major binding interactions in the inclusion complexes formed in noncompetitive solvents (CDCl₃ or CDCl₂CDCl₂)

tains an open, noncollapsed cavity with H-bonding sites in a helically-chiral orientation as revealed by ¹H-NMR analysis and molecular modeling using MacroModel (Mohamadi et al. 1990). Stoichiometric 1:1 host-guest association with *N*-Cbz-protected (Cbz=carbobenzyloxy) L- and D-Glu (Glu=glutamic acid) occurred in the noncompetitive solvent CDCl₂CDCl₂, and ¹H-NMR binding titration data (300 K) revealed that the optically active receptor bound *N*-Cbz-L-Glu (ΔG°=–3.2 kcal mol⁻¹) preferentially over *N*-Cbz-D-Glu (ΔG°=–2.1 kcal mol⁻¹). With a difference in stability between the two diastereoisomeric complexes of $\Delta(\Delta G^\circ)$=1.1 kcal mol⁻¹, (*S,S,S*)-(+)-**1** is a remarkably selective receptor given the rather moderate binding free enthalpy. This high selectivity is due to the cage-like architecture of the system, which creates specific size and shape requirements for guests and imposes specific constraints for the three-dimensional orientation of their H-bonding groups. According to ¹H-NMR analysis and computer modeling, the major binding mode in the resultant complexes formed is bidentate H-bonding of the spacer arms of the receptor to the COOH moieties of the substrate which occupies the cavity (Fig. 1b).

A first series of cleft-type receptors for the enantioselective recognition of excitatory amino acid derivatives incorporated a 9,9'-spirobi[9*H*-fluorene] spacer to which H-bonding sites (Garcia-Tellado et al. 1990) were attached (Cuntze et al. 1995). Compounds **2** and **3** (Fig. 2) formed

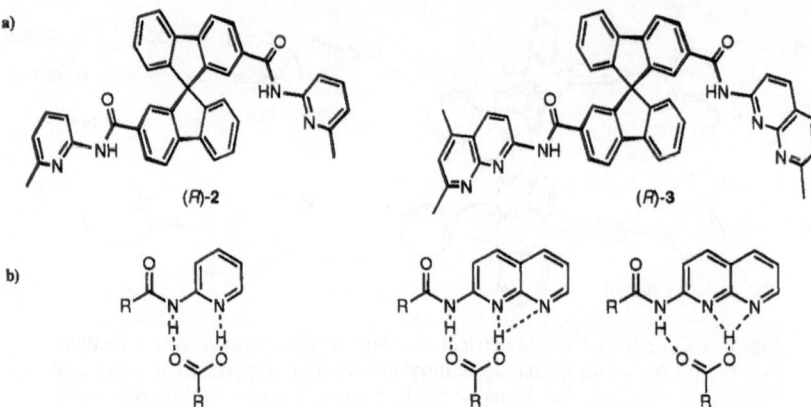

Fig. 2. a Cleft-type optically active receptors derived from 9,9'-spirobi[9H-fluorene] spacers, and **b** H-bonding patterns between a COOH residue and pyridinecarboxamides [as in (R)-2] or naphthyridinecarboxamides [as in (R)-3]

stable 1:1 complexes with N-protected Glu and Asp in CDCl₃, in which the major intermolecular interactions are the H-bonding between the two COOH residues of the substrate and the two heterocyclic carboxamides of the receptor as shown in Fig. 2b. Enantioselectivities of $\Delta(\Delta G^o)$ up to 0.9 kcal mol⁻¹ were measured with both receptors by ¹H-NMR titrations (Table 1). The relative orientation of the two COOH groups in N-Cbz-Glu with an odd-membered C-chain differs strongly from that in N-Cbz-Asp with an even-membered C-chain. As a result, the configuration of hosts **2** and **3** showing the strongest binding is reversed: N-Cbz-L-Glu is preferentially bound by the (S)-configured receptors whereas N-Cbz-L-Asp binds preferentially to the (R)-configured molecular clefts. In this article, we assign the absolute configuration by the nomenclature developed for spiro compounds whereas in the original paper (Cuntze et al. 1995) we treated **2** and **3** as compounds with axial chirality; these two alternative methods lead to opposite configurational assignments (Eliel et al. 1993).

Changing the H-bonding sites from pyridinecarboxamide (CONH(py)) in **2** to naphthyridinecarboxamide (CONH(naphthy)) in **3** did not significantly alter the free enthalpy and enantioselectivity of complexation due to two compensating effects. Binding to cleft **3** should

Table 1. Association constants K_a and binding free enthalpies ΔG°[a] determined by ^{1}H-NMR titrations[b] for the complexes formed by the enantiomers of **2** (at 293 K) and **3** (at 300 K) with N-Cbz-L-Asp and N-Cbz-L-Glu

Receptor	Substrate	K_a (l mol^{-1})	ΔG°(kcal mol^{-1})
(S)-**2**	N-Cbz-L-Asp	820	−3.9
(R)-**2**	N-Cbz-L-Asp	4200	−4.8
(S)-**3**	N-Cbz-L-Asp	1100	−4.2
(R)-**3**	N-Cbz-L-Asp	2600	−4.7
(S)-**2**	N-Cbz-L-Glu	14,000	−5.6
(R)-**2**	N-Cbz-L-Glu	3900	−4.8
(S)-**3**	N-Cbz-L-Glu	11,650	−5.6
(R)-**3**	N-Cbz-L-Glu	3800	−4.9

[a]Uncertainties: ±0.2 kcal mol^{-1}.
[b]In titrations at constant receptor concentration, the complexation-induced downfield shifts of the NH- and aromatic protons of the receptor were monitored and evaluated.

be weakened, since the pK_a value (in H_2O) of naphthyridine (pK_a=3.39) is significantly lower than of pyridine (pK_a=5.23), which makes the naphthyridine N-atoms weaker H-bond acceptors. On the other hand, binding to **3** should be strengthened as a result of a more favorable H-bonding pattern (DAA/AD as opposed to DA/AD; A=H-bond acceptor, D=H-bond donor), which should enable the formation of a bifurcated H-bond between the naphthyridine donor sites and the COOH proton (Fig. 2b). In addition, the DAA/AD pattern should also be more favorable in terms of secondary electrostatic interactions (Jorgensen and Pranata 1990; Zimmerman and Murray 1994).

Variable temperature ^{1}H-NMR studies showed that the complexation of the excitatory amino acid derivatives by **2** and **3** is strongly enthalpy-driven with the enthalpic driving force being partially compensated by an unfavorable change in entropy.

In a logical extension of the molecular recognition studies in the liquid phase, the optically active cleft (S)-**3** was covalently bound to silica gel to provide the new chiral stationary phase (CSP) (S)-**4** (Fig. 3; Cuntze and Diederich 1997). With CH_2Cl_2/MeOH 90:10 as the eluent, the enantiomers of (±)-N-Cbz-Glu were separable with a separation factor α=1.18. The L-enantiomer was eluted last, which correlates with

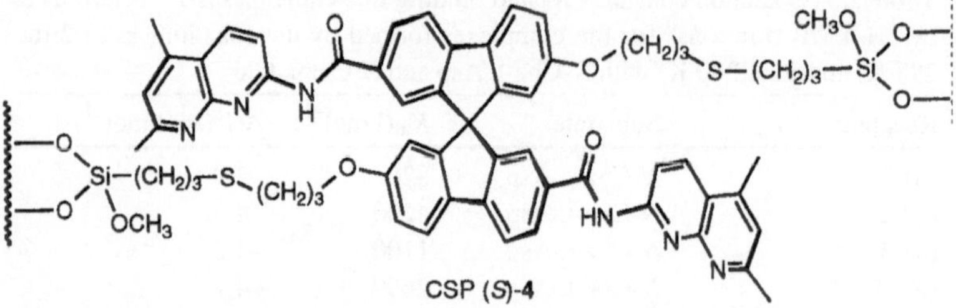

Fig. 3. A novel rationally designed chiral stationary phase (*CSP*) for the HPLC separation of excitatory amino acids and other dicarboxylic acids and diols

the results in the liquid phase where this enantiomer is bound more strongly by (*S*)-**3** (Table 1). Other dicarboxylic acids as well as a series of 1,1'-binaphthalene-2,2'-diols could also be separated on this new CSP. It is safe to expect that, with further refinement of the understanding of weak intermolecular interactions and rational receptor design, a great diversity of novel made-to-order chiral stationary phases (Gasparrini et al. 1995) will become commercially available. This development should, therefore, benefit the pharmaceutical industry which is required to produce chiral drugs in enantiomerically pure form only.

Whereas the spirobifluorene clefts (*R*)-**2** and (*S*)-**2** displayed efficient chiral recognition in the complexation of *N*-Cbz-L-Glu or *N*-Cbz-L-Asp (Table 1), the corresponding 1,1'-binaphthalene-based receptors (*R*)-**5** and (*S*)-**5** (Fig. 4) formed stable diastereoisomeric complexes of nearly identical association strength with the excitatory amino acid derivatives (Table 2; Alcázar and Diederich 1992). These experimental findings provide impressive support for the hypothesis that a high degree of receptor preorganization and conformational homogeneity is a requirement for efficient chiral recognition. In contrast to the rigid spirobifluorene cleft, the 1,1'-binaphthyl unit is conformationally flexible and capable of adopting geometries which fit both substrate enantiomers. As a consequence, diastereoisomeric complexes of similar energy are formed.

In subsequent work, the 1,1'-binaphthalene-derived receptors were incrementally improved and these developments have ultimately led to strong binders which discriminate efficiently [enantioselectivities

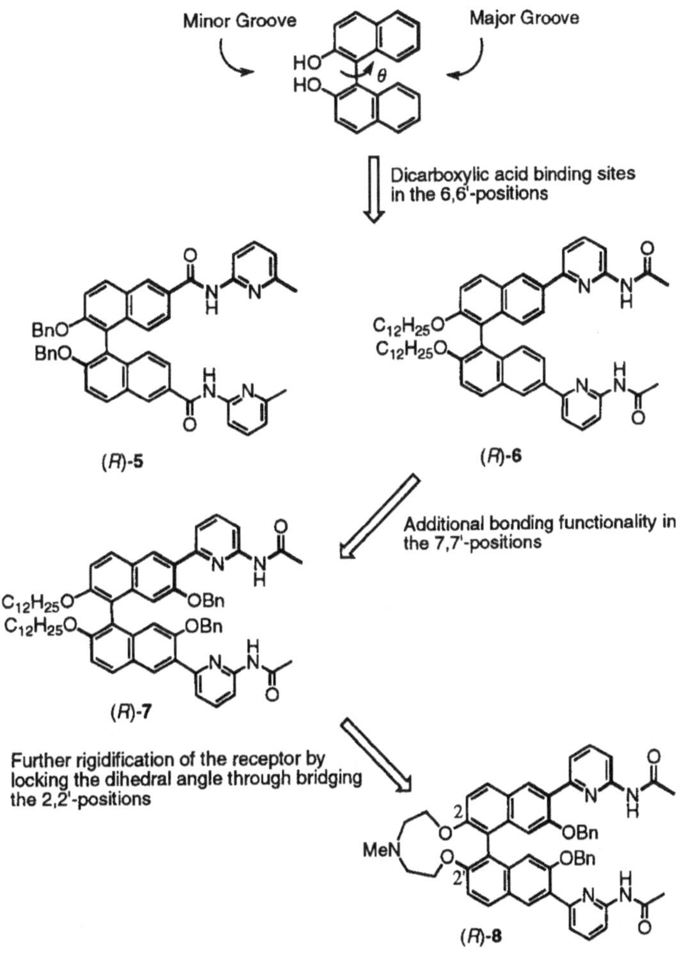

Fig. 4. Sequential optimization of 1,1'-binaphthalene receptors for the enantioselective complexation of derivatives of excitatory amino acids. Bn=PhCH$_2$

Table 2. Association constants K_a, binding free enthalpies ΔG^{oa}, and enantiose-lectivities $\Delta(\Delta G^o)$ determined by ^{1}H-NMR titrations[b] for the complexes formed by 1,1'-binaphthalene-derived receptors with N-Cbz-Asp and N-Cbz-Glu (293–300 K)

Receptor	Substrate	K_a (l mol^{-1})	ΔG^o (kcal mol^{-1})	$\Delta(\Delta G^o)$ (kcal mol^{-1})
(R)-5	N-Cbz-L-Asp	2000	–4.5	
(S)-5	N-Cbz-L-Asp	3300	–4.8	0.3
(R)-5	N-Cbz-L-Glu	20,800	–5.8	0.1
(S)-5	N-Cbz-L-Glu	19,400	–5.7	
(R)-6	N-Cbz-L-Asp	4000	–4.9	
(S)-6	N-Cbz-L-Asp	8900	–5.4	0.5
(R)-6	N-Cbz-L-Glu	3000	–4.7	0.3
(S)-6	N-Cbz-L-Glu	1800	–4.4	
(R)-7	N-Cbz-L-Asp	180,000	–7.2	0.7
(S)-7	N-Cbz-L-Asp	60,000	–6.5	
(R)-7	N-Cbz-L-Glu	19,000	–5.8	0.3
(S)-7	N-Cbz-L-Glu	11,000	–5.5	
(R)-8	N-Cbz-L-Asp	87,000	–6.80	1.65
(R)-8	N-Cbz-D-Asp	5600	–5.15	
(R)-8	N-Cbz-L-Glu	94,000	–6.85	1.25
(R)-8	N-Cbz-D-Glu	12,000	–5.60	

[a]Uncertainties: ±0.2 kcal mol^{-1}.

[b]In titrations at constant receptor concentration, the complexation-induced downfield shifts of the NH- and aromatic protons of the receptor were monitored and evaluated.

$\Delta(\Delta G^o)$ up to 1.65 kcal mol^{-1}] between the enantiomers of the excitatory amino acid derivatives. Firstly, the orientation of the pyridinecarboxamide binding sites in (R)-5 and (S)-5 was changed to yield receptors (R)-6 and (S)-6, in which the pyridine rings had been directly coupled to the 1,1'-binaphthalene via the Suzuki cross-coupling reaction (Martinborough et al. 1995; Fig. 4). Although the overall binding free enthalpy was decreased, a modest enhancement in chiral recognition of N-Cbz-L-Glu was observed (Table 2). Subsequently, additional functional groups were introduced into the 7,7'-positions of the 1,1'-binaphthalene cleft. The introduction of two benzyloxy residues in (R)-7 and (S)-7 strongly enhanced the overall binding affinity, in particular for

N-Cbz-Asp ($-\Delta G^0$ up to 7.2 kcal mol^{-1}), and also provided an improved degree of chiral recognition [up to $\Delta(\Delta G^0)$=0.7 kcal mol^{-1}]. The benzyl ether functionality in the 7,7'-positions sterically forces the pyridine rings out of the planes of the adjacent naphthalene rings and enforces conformations in which the CONH(py) binding sites converge into the cleft for interaction with the dicarboxylic acid guests. Furthermore, the benzyl ether O-atoms are sufficiently close to the OH-groups of the substrates binding to the adjacent pyridine N-atoms, so that additional attractive electrostatic O···H interactions with these OH groups become effective.

Finally, locking the dihedral angle θ about the C(1)-C(1')-axis of the 1,1'-binaphthalene spacer (Fig. 4) by bridging the 2,2'-positions led to a strongly enhanced receptor preorganization and, consequently, a dramatic increase in enantioselectivity. A series of receptors with locked dihedral angles θ varying between 56° and 102°was rationally designed and subsequently prepared (Lustenberger et al. 1998). The highest enantioselectivity in the recognition of excitatory amino acid derivatives in CDCl$_3$ [N-Cbz-Asp: $\Delta(\Delta G^0)$=1.65 kcal mol^{-1}; N-Cbz-Glu: $\Delta(\Delta G^0)$=1.25 kcal mol^{-1}; 300 K] was reached with (R)-8 (Fig. 4) with the dihedral angle locked at θ=86±4°(Table 2).

The more stable diastereoisomeric complexes are more highly structured in solution than their less stable counterparts. This was clearly demonstrated by ^{1}H{^{1}H} nuclear Overhauser effect (NOE) difference spectroscopy. A total of 5 intermolecular NOEs were measured for the complex between (R)-8 and N-Cbz-L-Asp (Fig. 5), whereas only two intermolecular NOEs were observed for the diastereoisomeric complex formed by N-Cbz-D-Asp. Computer simulation of the more stable complex (MacroModel V. 5.0, OPLS* force field, 1000-step Monte Carlo multiple miminum search in CDCl$_3$) yielded a total of 4 structures within 2.4 kcal mol^{-1} of the computed global minimum (Fig. 5), which were all congruent with the experimentally observed NOEs. The torsional angles between the pyridyl and adjacent naphthalene moieties of the receptor were found to be 70°, and the dihedral angle θ was computed to be ca. 86°. The substrate binds in the *cis*-carbamate conformation. Characteristic host-guest interactions in the tight complex are the two COOH···CONH(py) H-bonding motifs with nearly perfectly linear O-H···N H-bonds of ca. 1.9 Å and C=O···H-N H-bonds of ca. 1.7 Å length. The bound COOH groups further participate in attractive secon-

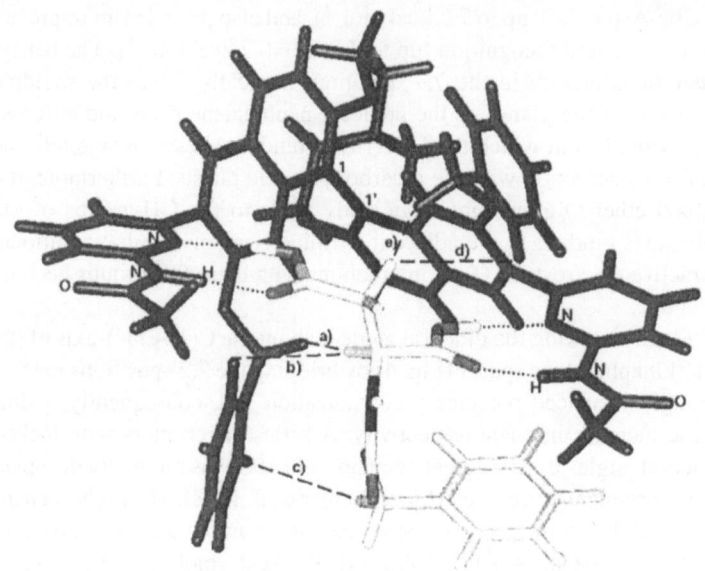

Fig. 5. Computer-generated model of the complex formed between (*R*)-**8** and *N*-Cbz-L-Asp in CDCl₃. The four COOH···CONH(py) H-bonds (........) are shown as well as the five observed intermolecular NOEs (– – – –) at computed distances of 2.39 Å (a), 2.83 Å (b), 3.25 Å (c), 3.26 Å (d), and 2.53 Å (e)

dary electrostatic H-bonding interactions C(O)O-H···O-C(7), with O···O distances between 3.3 and 3.4 Å and O-H···O angles of 105°. Aromatic interactions also play an important role in the stabilization of the complex: one phenyl ring of the receptor stacks with the *cis*-carbamate bond of the substrate, both phenyl rings of the receptor participate in T-shaped interactions with the two COOH···CONH(pyr) H-bond arrays (Muehldorf et al. 1988), and the phenyl ring of the substrate is involved in a face-to-face stacking with one of these host-guest H-bonding motifs. A much poorer host-guest complementarity with a smaller number of intermolecular contacts was found in the computed structures of the less stable diastereoisomeric complex between (*R*)-**8** and *N*-Cbz-D-Asp, for which only two NOEs were observed experimentally. Furthermore, the two COOH···CONH(pyr) H-bonding arrays in this complex are strongly distorted out of planarity.

Enforcing conformational homogeneity by bridging represents a new general principle for improving the chiral recognition potential of 1,1'-binaphthyl receptors and, in principle, other such flexible receptor systems also.

3.3 Carbohydrate Recognition

Given the complex nature of the interactions between proteins and carbohydrates displayed in an increasing number of X-ray crystal structures (Quiocho 1989; Sharff et al. 1995), it has become apparent to both chemists and biologists that many of the underlying principles governing these interactions cannot be identified or quantified on an atomic scale in biological studies (Bourne et al. 1993; Lemieux 1991). Rather, there is increasing consensus that investigations with well-defined synthetic receptors (Bonar-Law and Sanders 1995), whose binding properties can be systematically varied and analyzed, will make important contributions to the understanding of carbohydrate recognition processes.

The successful complexation of sugars in water is based on a subtle balance between hydrophobic and hydrophilic interactions. Highly directional, often ionic, bidentate H-bonds (to Asp, Glu, Arg, Asn, and Gln residues of proteins) control the binding selectivity, and hydrophobic forces provide the thermodynamic driving force for complexation. There are many remaining questions, however, which systematic studies with synthetic receptors could answer: How many H-bonds are needed to form a stable carbohydrate complex in water? In the complex between bacterial arabinose binding protein and L-arabinose, as many as 21 H-bonds are observed, yet the binding free enthalpy of such complexes amounts only to values of $-\Delta G=7$–8 kcal mol^{-1} (Vermersch et al. 1991). Which ionic H-bonding residues (anionic Asp, Glu, phosphates vs cationic Arg, His) provide the strongest interaction with sugars? What is the contribution of cooperativity (i.e., a sugar OH-group acts both as a H-bond donor and acceptor) to H-bonding strength? How important are apolar interactions such as van der Waals dispersion interactions, hydrophobic desolvation, and sugar-CH···aromatic π-electron interactions? The biological X-ray crystal structures consistently show stacking between aromatic amino acid side chains, predominantly Tyr and

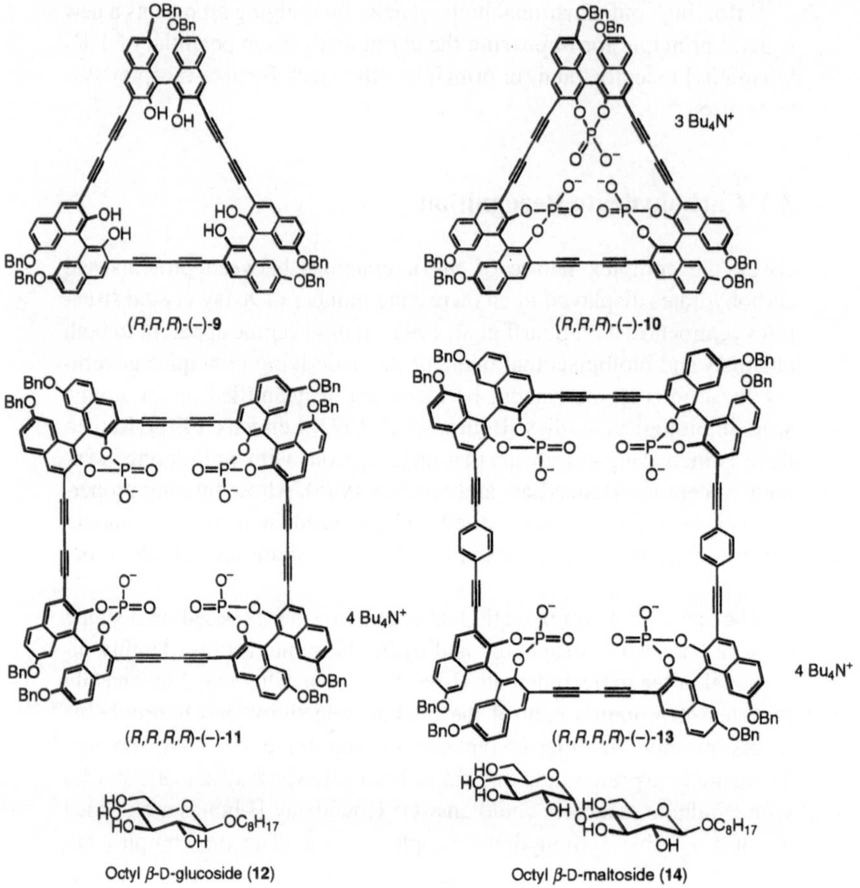

Fig. 6. Carbohydrate binding receptors and their substrates

Trp, with the apolar faces of the carbohydrates. A great variety of medicinal chemistry programs would benefit from sound answers to these questions (Wong et al. 1995).

To study carbohydrate recognition, we prepared a series of optically active cyclophane receptors (Fig. 6). By macrocyclic bridging of 1,1'-binaphthalene moieties in the 3,3'-positions via rigid acetylenic linkers, H-bonding groups attached to the 2,2'-positions converge into a highly preorganized recognition site (Anderson et al. 1995). ^{1}H-NMR binding titrations showed that (R,R,R)-(–)-**9**, with six neutral convergent H-bonding centers, formed moderately stable 1:1 complexes with 1-octyl pyranosides in dry CDCl$_3$, with free enthalpies of formation $-\Delta G^o$ between 2.7 and 3.5 kcal mol^{-1} (association constants K_a between 90 and 370 l mol^{-1}, 300 K). Under the same conditions, no significant complexation of pyranosides was observed in control runs with a single one of the three dialkynylated 1,1'-binaphthalene-2,2'-diol moieties present in (R,R,R)-(–)-**9**. Slow host-guest exchange kinetics near the NMR time scale provided support for a complex geometry in which the sugar substrate fully penetrates into the receptor cavity rather than docking onto one of the two receptor faces.

In order to mimic the numerous ionic bidentate H-bonding motifs seen in the X-ray crystal structures of protein-saccharide complexes, the complexation between the trianionic macrocycle (R,R,R)-(–)-**10** and monosacharides was explored. However, the large convergent phosphodiester groups greatly reduced the open space in the inner cavity and forced carbohydrates to bind outside the cavity onto one of the receptor faces in a docking mode. Moderately strong complexation between (R,R,R)-(–)-**10** and 1-octyl pyranosides (K_a=2500–3000 l mol^{-1}) was observed in CD$_3$CN which, however, vanished upon addition of small amounts of competitive protic co-solvent.

An impressive increase in performance was achieved by organizing multiple ionic H-bonding groups in a cavity complementary to the size of a pyranoside. The tetraanionic receptor (R,R,R,R)-(–)-**11** complexed 1-octyl β-D-glucoside **12** (and other monopyranosides) in highly competitive solvent mixtures such as CD$_3$CN containing up to 20% v/v CD$_3$OD as protic co-solvent. A high association constant K_a of 5200 l mol^{-1} (ΔG^o=–5.1 kcal mol^{-1}) was measured for the 1:1 complex with **12** in 2% CD$_3$OD/CD$_3$CN, a competitive solvent mixture which could not be used in previous carbohydrate recognition studies (Bonar-Law and Sanders 1995; Bhattarai et al. 1992).

By introducing *p*-phenylene spacer groups within two parallel acetylenic linkers (Neidlein and Diederich 1996), an expanded binding site for the selective incorporation of disaccharides was created. Thus, (R,R,R,R)-(–)-**13** formed a highly stable complex with octyl β-D-maltoside (**14**) (as well as with the corresponding melibioside and lactoside) in the competitive solvent mixture 12% CD$_3$OD/CD$_3$CN (K_a=11,000 l mol^{-1}, ΔG°=−5.5 kcal mol^{-1}, 300 K). The association strength was evaluated by ^{1}H-NMR binding titrations in which, at varying receptor concentration, the complexation-induced change in chemical shift of the anomeric proton H-C(1) of the disaccharide was monitored. The selectivity of (R,R,R,R)-(−)-**13** for disaccharides over monosaccharides in the protic solvent mixture is impressive and, from comparative studies, must be estimated as $\Delta(\Delta G^\circ)$>3 kcal mol^{-1}. This high selectivity is readily explained by the size of the cavity of (R,R,R,R)-(−)-**13**, which fits disaccharides well but, unlike the cavity of (R,R,R,R)-(−)-**11**, is much too spacious for incorporating a monosaccharide with formation of ionic H-bonds to all four convergent phosphates. In a nonprotic solvent (CD$_3$CN), (R,R,R,R)-(−)-**13** was found to bind monosaccharide **12**, but with a host-guest stoichiometry higher than 1:1, and presumably 2:1 as suggested by *Job* plot analysis.

In further developments, the phenyl rings of (R,R,R,R)-(−)-**13** will be functionalized to provide additional recognition sites as well as catalytic sites for cleavage of the glycosidic bond in complexed disaccharides. Furthermore, they provide anchor points for attachment of additional, capping aromatic rings which should undergo apolar interactions (dispersion interactions, hydrophobic desolvation) with the bound sugar, ultimately enabling complexation in aqueous solution.

3.4 Double-Decker Cyclophane Receptors for Steroids: Dissolution of Cholesterol in Water

The three chiral double-decker cyclophane receptors for steroidal substrates D_2-(±)-**15**, D_2-(±)-**16**, and C_2-(±)-**17** (Fig. 7; Marti et al. 1998) were prepared by synthetic sequences in which key C–C bond forming reactions were accomplished by Pd(0)-catalyzed cross-coupling reactions such as the *Stille* (Mitchell 1997) and *Hiyama* couplings (Hiyama 1997). The smaller cylindrical receptor (±)-**15** with ethynediyl bridges possesses an 11-Å deep and 8×11-Å wide cavity (Peterson and Diederich 1994), while cylindrical (±)-**16** with buta-1,3-diynediyl bridges provides an enlarged pocket of 9×12 Å width and 13 Å depth (Peterson et al. 1995). With both receptors, dissolution of solid cholesterol (**18**, Fig. 8) in water with the formation of a stable 1:1 host-guest complex was achieved. A 1 mM solution of the more efficient cholesterol binder

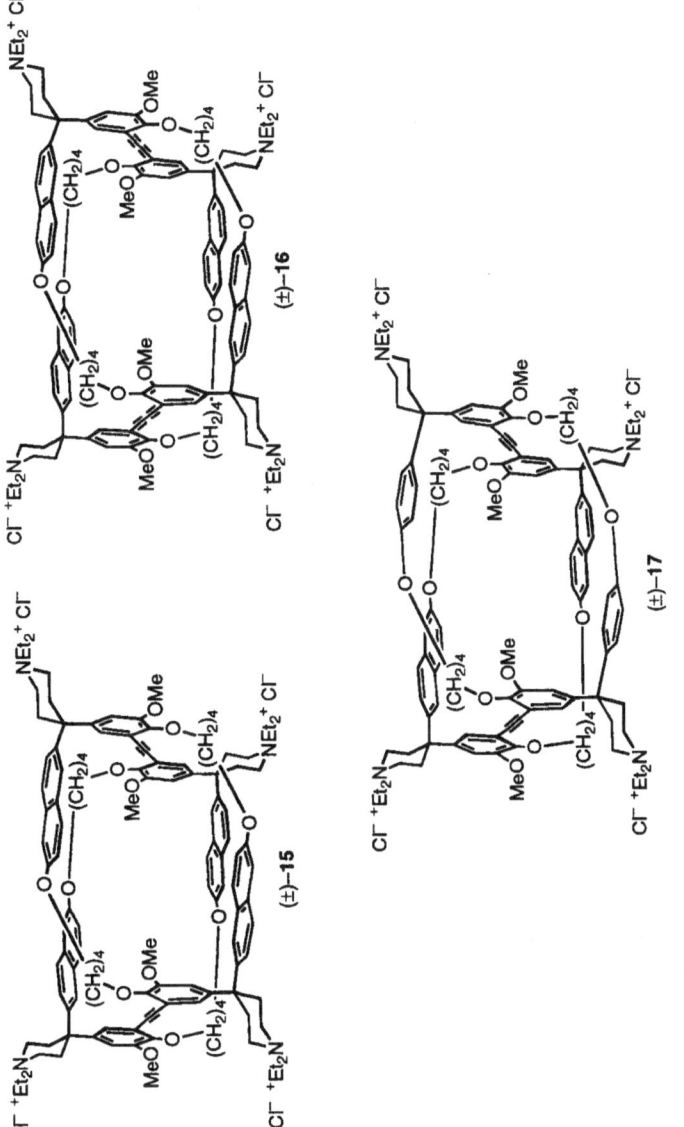

Fig. 7. Double-decker cyclophane receptors for steroid complexation

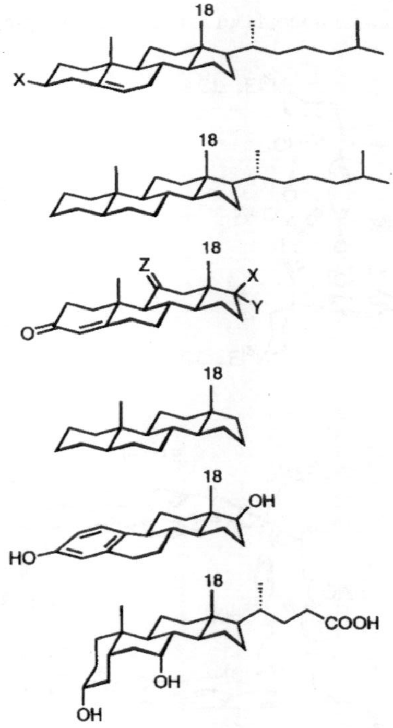

		X
19	cholesteryl acetate	OCOCH₃
20	5-cholestene	H
18	cholesterol	OH

| 21 | 5α-cholestane |

		X	Y	Z
22	testosterone	OH	H	H₂
26	cortisone	COCH₂OH	OH	O
27	hydrocortisone	COCH₂OH	OH	α-H, β-OH

| 23 | 5α-androstane |

| 24 | β-estradiol |

| 25 | chenodeoxycholic acid |

Fig. 8. Some of the steroids investigated by ^{1}H-NMR binding studies with receptors (±)-**15**, (±)-**16**, and (±)-**17** in CD₃OD

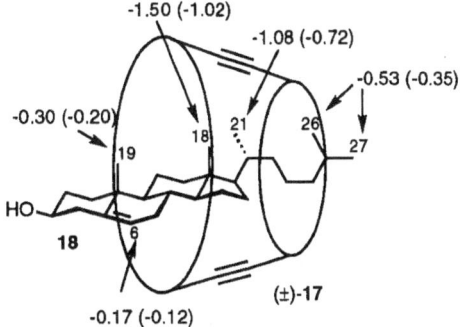

Fig. 9. Schematic representation of the preferred binding mode of cholesterol (**18**) by the conical receptor (±)-**17** based on the depicted complexation-induced changes in chemical shift at saturation binding ($\Delta\delta_{max}$) and the maximum shifts ($\Delta\delta_{max\ obs}$; *shown in parenthesis*) observed in ^{1}H-NMR binding titrations at constant steroid concentration

(±)-**15** enhanced the solubility of the steroid by a factor of 180 from 4.7 μM to 0.85 mM. From these solid-liquid extraction data, the association constant for the formed 1:1 complex was calculated as $K_a=1.1\times10^6$ l mol^{-1} ($\Delta G^o=-8.2$ kcal mol^{-1}). This was the first observation of efficient cholesterol dissolution in water by a synthetic receptor.

To elucidate the principles governing selective molecular recognition of steroids, comparative ^{1}H-NMR titration binding studies were undertaken in CD$_3$OD with the two cylindrical receptors (±)-**15** and (±)-**16** and a diversity of steroidal substrates (Table 3). In the 1:1 complexes formed, the steroid is axially incorporated in the cavity, i.e., the longest axis extends in the direction of the C_2-axis of the receptors. All four rings A, B, C, and D are fully incorporated in the binding site. Some of the most striking observations are summarized in the following: (a) The different binding characteristics of fully aliphatic and flatter, partially unsaturated steroids were clearly revealed. Cholesterol (**18**) and derivatives such as cholesteryl acetate (**19**) or 5-cholestene (**20**), with a 5,6-double bond, are preferably incorporated into the shallower cavity of (±)-**15**, whereas fully aliphatic derivatives such as 5α-cholestane (**21**) prefer complexation in the deeper and wider cavity of (±)-**16**. These experimental steroid binding selectivities could be reproduced in computer simulations (Peterson et al. 1995). The difference in cavity width

Table 3. Association constants K_a and binding free enthalpies ΔG^o from ^{1}H-NMR titrations for 1:1 steroid complexes formed by receptors (±)-**15** and (±)-**16** in CD$_3$OD at 298 K[a]. Also shown are the maximum observed complexation-induced upfield shifts $\Delta\delta_{max\ obs}$ and the upfield shifts at saturation binding $\Delta\delta_{sat}$ of the steroidal Me(18) resonance, which was evaluated in the titrations

Steroid		Receptor (±)-**15**			Receptor (±)-**16**		
		K_a (mol l^{-1})	ΔG^o (kcal mol^{-1})	$\Delta\delta_{max\ obs}$ ($\Delta\delta_{sat}$) Me(18)	K_a (mol l^{-1})	ΔG^o (kcal mol^{-1})	$\Delta\delta_{max\ obs}$ ($\Delta\delta_{sat}$) Me(18)
19	Cholesteryl acetate	4800	-5.0	-1.54 (-1.95)	2300	-4.6	-0.95 (-1.33)
20	5-Cholestene	3200	-4.8	-0.85 (-1.19)	2300	-4.6	-0.85 (-1.20)
18	Cholesterol	1500	-4.3	-1.14 (-1.70)	900	-4.1	-0.64 (-0.97)
21	5α-Cholestane	870	-4.0	-0.90 (-1.57)	2700	-4.7	-0.81 (-1.10)
22	Testosterone	2100	-4.5	-1.25 (-1.69)	200	-3.1	-0.34 (-1.13)
26	Cortisone	160	-3.0	-0.36 (-1.52)	Not determined		
27	Hydrocortisone	110	-2.8	-0.13 (-0.60)	Not determined		
23	5α-Androstane	500	-3.7	-0.72 (-1.74)	370	-3.5	-0.51 (-1.16)
24	β-Estradiol	390	-3.5	-0.81 (-2.04)	170	-3.0	-0.29 (-1.10)
25	Chenodeoxycholic acid	40	-2.2[b]		Not determined		

[a]The resonances of Me(18) in the two possible diastereoisomeric complexes between racemic receptor and enantiomerically pure steroid were not resolved, and the association constants give an average value for both complexes. Reproducibility of K_a values: ± 10%.
[b]Estimated upper limit.

between the two receptors also leads to a striking difference in binding affinity for testosterone (**22**) with its flat A-ring. Whereas (±)-**15** complexes testosterone strongly ($\Delta G^o=-4.5$ kcal mol^{-1}), a much weaker 1:1 complex ($\Delta G^o=-3.1$ kcal mol^{-1}) is formed by (±)-**16** with its wider cavity. (b) Unprecedented information on the contributions from complexation of the isoprenoidal side chain of the steroidal substrates to the overall stability of the inclusion complexes was obtained. For instance, 5α-cholestane (**21**) binds much more strongly [$\Delta(\Delta G^o)=1.2$ kcal mol^{-1}] to receptor (±)-**16**, with a 13-Å-deep binding site, than 5α-androstane, (**23**) which lacks the isoprenoidal side chain (Table 3). Diagnostic upfield complexation-induced shifts of the Me-resonances in the side chain of **21** provided experimental evidence for the additional incorporation of the flexible chain of the guest in the host cavity. In contrast, the shallower (11 Å) cavity of receptor (±)-**15** does not incorporate the isoprenoidal side chain efficiently and 5α-cholestane (**21**) and 5α-androstane (**23**) formed complexes of similar stabilities. Similarly, ^{1}H-NMR titrations in CD$_3$OD (298 K) revealed that the conical receptor (±)-**17** incorporates the steroidal side-chain of cholesterol (**18**; $\Delta G^o=-3.8$ kcal mol^{-1}; $K_a=600$ l mol^{-1}) into the narrowing section of its cavity, whereas the cholesterol A-ring protrudes into the solvent (Fig. 9; Marti et al. 1998). (c) Both receptors (±)-**15** and (±)-**16** discriminate between aliphatic and aromatic guests, the complexes with β-estradiol (**24**) being weak in comparison with those of most aliphatic guests. Also, bile acids form weak complexes with (±)-**15**, presumably due to the less favorable *cis*-configuration of their A-B rings. Chenodeoxycholic acid (**25**) undergoes a particularly weak association since its OH-group at C(7) will become desolvated upon incorporation into the hydrophobic receptor cavity. Unfavorable functional group desolvation (Carcanague and Diederich 1990) upon incorporation into the host cavity might also be the origin of the weak complexation of cortisone (**26**) and hydrocortisone (**27**) by (±)-**15**.

   The structural data obtained from the ^{1}H-NMR binding titrations with (±)-**15**, (±)-**16**, and (±)-**17** illustrate a distinct advantage of cyclophane receptors over cyclodextrins: As a result of the anisotropic effects of the aromatic cavity walls on the chemical shifts of protons in bound steroids, molecular recognition studies with cyclophane receptors provide extensive information on the geometries of the inclusion complexes formed. In contrast, cyclodextrin receptors without aromatic rings affect the chemical shifts of protons in bound substrates only weakly and structural information therefore is limited.

3.5 Dendrophanes: Dendritic Receptors for the Complexation of Arenes and Steroids in Aqueous Solution

The dendrophanes (*dendr*itic cyclo*phanes*) of first [28, $C_{102}H_{144}N_4O_{44}$, molecular weight (MW) 2130 daltons (D)], second (29, $C_{258}H_{396}N_{16}O_{140}$, MW 5962), and third generation (30, $C_{726}H_{1152}N_{52}O_{478}$, MW 17457) (Fig. 10) form 1:1 host-guest complexes with benzene and naphthalene derivatives such as **31–33** in basic aqueous buffer containing small amounts of organic co-solvent (Table 4; Mattei et al. 1995). With their larger cyclophane initiator core, dendrophanes of first to third generation **34** ($C_{126}H_{148}N_4O_{48}$, MW 2487 D), **35** ($C_{282}H_{400}N_{16}O_{144}$, MW 6318 D), and **36** ($C_{750}H_{1156}N_{52}O_{432}$, MW 17813) (Fig. 11) bind testosterone (**22**) in basic aqueous buffer/MeOH 1:1 (Wallimann et al. 1996). The third generation compounds **30** and **36** are unprecedented models for binding sites deeply buried in globular proteins; computer modeling revealed that the central cyclophane is completely encapsulated by the surrounding dendritic branches.

A combination of ^{1}H-NMR and fluorescence binding titrations (Table 4) provided the following unprecedented results (Mattei et al. 1997): (a) The cyclophane recognition sites at the center of the dendrophanes remain open and effective at all dendritic generations studied. Thus, they form inclusion complexes of similar stability to those formed by the corresponding initiator core cyclophanes (not shown). In all complexes, the substrates are exclusively located in the central cyclophane cavities and nonspecific incorporation into fluctuating voids in the dendritic shell is negligible. (b) Fluorescence titrations with the fluorescent probe 6-(*p*-toluidino)naphthalene-2-sulfonate (TNS, **33**) demonstrated that the micropolarity around the binding cavity in the series **28–30** became significantly reduced with increasing dendritic superstructure. The micropolarity at the center of the third-generation dendrophane **30** in water is comparable to that of EtOH. (c) The host-guest exchange kinetics for all dendrophanes is remarkably fast: ^{1}H-NMR binding titrations, which rely on fast host-guest exchange ($k_{decompl} > 10^2$ to 10^3 s^{-1}), were possible with all dendrophanes except **30**. The fast host-guest exchange kinetics

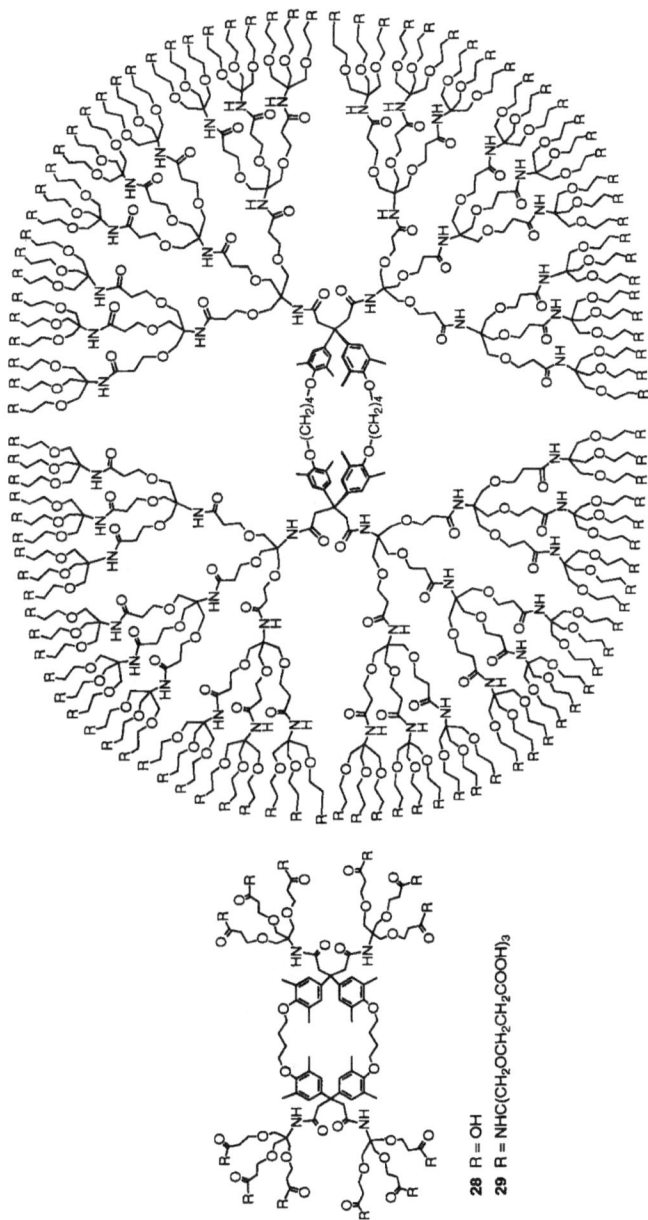

30 R = COOH

28 R = OH
29 R = NHC(CH₂OCH₂CH₂COOH)₃

Fig. 10. *Dendritic cyclophanes* (dendrophanes) **28–30** of first to third generation for the complexation of arenes in aqueous solution

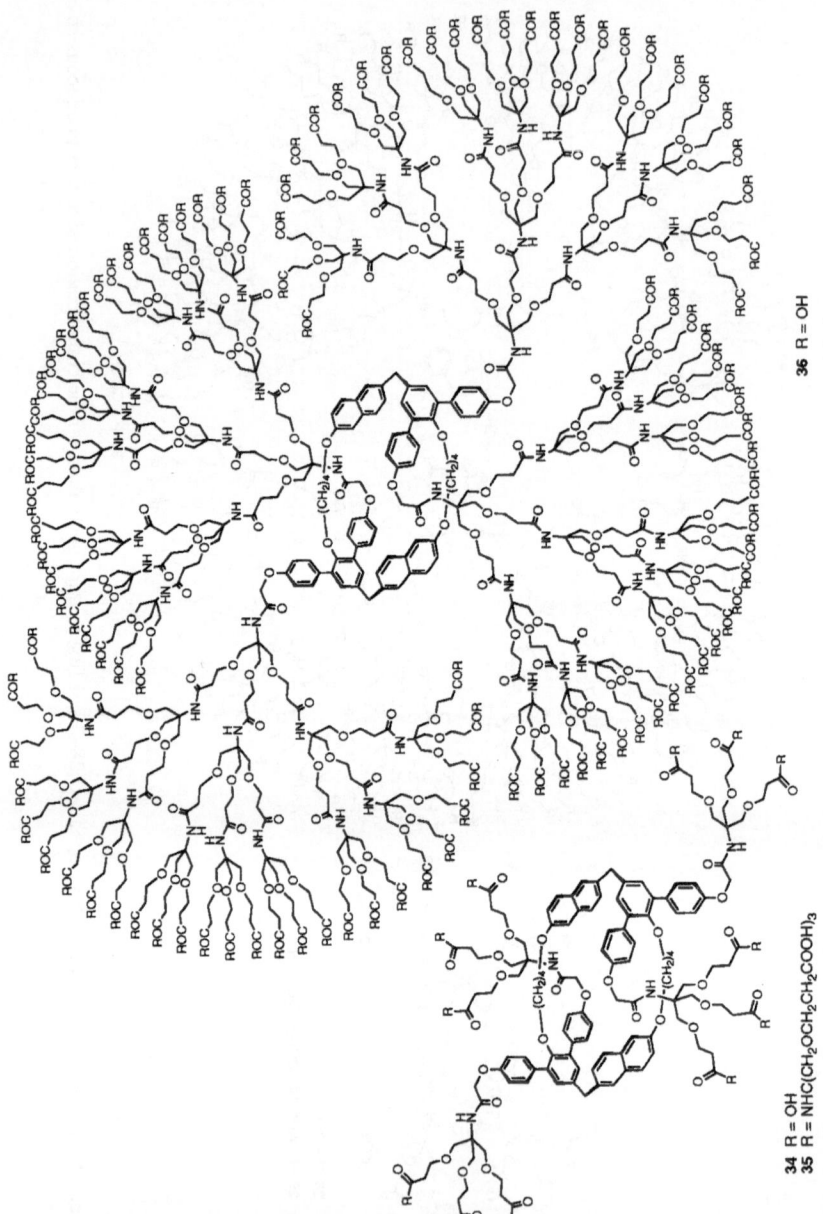

Fig. 11. Dendrophanes **34–36** of first to third generation for the complexation of steroids in aqueous solution

34 R = OH
35 R = NHC(CH₂OCH₂CH₂COOH)₃

36 R = OH

Table 4. Association constants K_a (1 mol^{-1}) and complexation free enthalpies ΔG^o (kcal mol^{-1}) for dendrophane complexes in aqueous solutions (298–300 K)

Dendrophane	Guest	K_a (1 mol^{-1})	ΔG^o (kcal mol^{-1})[a]
^{1}H-NMR titrations in 0.066 M phosphate buffer in D$_2$O (pD 8.4)[b]			
28	**31**[c]	1800	–4.4
29	**31**[c]	1700	–4.3
28	**32**[c]	370	–3.5
28	**33**[d]	8000	–5.3
29	**33**[d]	2200	–4.6
Fluorescence titrations in 0.066 M phosphate buffer in H$_2$O (pH 8.0)[b]			
28	**33**	10,500	–5.5
29	**33**	8000	–5.3
30	**33**	5500	–5.1
^{1}H-NMR titrations in borate-buffered D$_2$O (pD 10.5)/CD$_3$OD 1:1[b]			
34	**22**	700	–3.9
35	**22**	750	–3.9
36	**22**	1100	–4.2

[a]Uncertainties in ΔG^o: ±0.1 kcal mol^{-1}.
[b]Guest concentration held constant.
[c]In D$_2$O buffer/(CD$_3$)$_2$SO 97.3:2.7.
[d]In D$_2$O buffer/CD$_3$OD 85:15.

with **36** was confirmed by fluorescence relaxation measurements with fluorescent steroidal substrates (Kühn-Velten and Kempfle 1993; M.A. Kempfle, P. Wallimann, F. Diederich, unpublished). This result is in sharp contrast to the findings by Meijer and coworkers (Jansen et al. 1995), who observed substrate encapsulation for hours and even days in their poly(propylene imine) dendrimers. These different findings can be readily explained: the Meijer dendrimers have a tight, densely packed superstructure diverging from a small initiator core. In contrast, the four dendritic wedges in **28–30** and **34–36** are attached to large, nanometer-sized cyclophane cores which produce apertures through which substrates can rapidly enter or leave the binding cavity. Furthermore, the Meijer-dendrimers possess both strongly H-bonding and sterically encumbering surface groups which generate tight substrate encapsulation at the interior whereas the carboxylates at the dendrophane surfaces will not densely pack for electrostatic reasons.

The reduced micropolarity at the central cyclophane core (Dandliker et al. 1997) and the fast host-guest exchange kinetics make H_2O-soluble dendrophanes attractive targets as catalytically active mimics of globular enzymes. Such *catalytic dendrophanes*, with catalytic cyclophane initiator cores that are shielded from the aqueous solution by dendritic superstructure, are now under construction, targeting the acceleration of reactions which particularly benefit from a reduced environmental polarity.

3.6 Conclusions

Molecular recognition studies with synthetic receptors nicely complement the investigations with biological receptors in enhancing the understanding of the principles governing selective complexation of small biomolecules. Since the principles of molecular recognition in chemical and biological systems are identical, the knowledge gained from studies such as those described in this article can be directly applied in medicinal chemistry to the rational, structure-based design of lead compounds or to lead optimization. Novel perspectives for biomedicinal technologies may open up: The development of highly specific, tight-binding steroid receptors could provide new strategies for interfering with biologically important steroids in vivo and potentially lead to a new class of pharmacological agents. Receptors specific for cholesterol may offer an alternative pharmacological strategy for the dissolution of cholesterol deposits such as those in atherosclerotic plaque. Furthermore, developments in analytic, sensoric, and separation technologies greatly benefit from an enhanced insight into molecular recognition principles. It is safe to predict that the field of molecular recognition will continue to prosper in the future in view of its relevance to so many key technologies in chemistry, biology, and materials sciences.

Acknowledgements. We thank the Chiral 2 program of the Swiss National Science Foundation, the ETH research council, and F. Hoffmann-La Roche, Basel, for their generous support of this work.

References

Alcázar V, Diederich F (1992) Enantioselective complexation of chiral dicarboxylic acids in functionalized 9,9'-spirobifluorene clefts. Angew Chem Int Ed Engl 31:1521–1523

Anderson S, Neidlein U, Gramlich V, Diederich F (1995) A new family of chiral binaphthyl-derived cyclophane receptors: complexation of pyranosides. Angew Chem Int Ed Engl 34:1596–1600

Arevalo JH, Stura EA, Taussig MJ, Wilson IA (1993) Three-dimensional structure of an anti-steroid Fab' and Progesterone-Fab' complex. J Mol Biol 231:103–118

Bhattarai KM, Bonar-Law RP, Davis AP, Murray BA (1992) Diastereo- and enantio-selective binding of octyl glucosides by an artificial receptor. J Chem Soc Chem Commun 752–754

Bonar-Law RP, Sanders JKM (1995) Polyol recognition by a steroid-capped porphyrin. Enhancement and modulation of misfit guest binding by added water or methanol. J Am Chem Soc 117:259–271, and references cited therein

Bourne Y, van Tilbeurgh H, Cambillau C (1993) Protein-carbohydrate interactions. Curr Opin Struct Biol 3:681–686

Carcanague DR, Diederich F (1990) A spacious cyclophane host for inclusion complexation of steroids and [m.n]paracylophanes. Angew Chem Int Ed Engl 29:769–771

Cuntze J, Diederich F (1997) Molecular recognition and enantiomer separations on a novel chiral stationary phase based on a 9,9'-spirobi[9H-fluorene]-derived molecular cleft. Helv Chim Acta 80:897–911

Cuntze J, Owens L, Alcázar V, Seiler P, Diederich F (1995) Molecular clefts derived from 9,9'-spirobi[9H-fluorene] for enantioselective complexation of pyranosides and dicarboxylic acids. Helv Chim Acta 78:367–390

Dandliker PJ, Diederich F, Zingg A, Gisselbrecht JP, Gross M, Louati A, Sanford E (1997) Dendrimers with porphyrin cores: synthetic models for globular heme proteins. Helv Chim Acta 80:1773–1801

Diederich F (1991) Cyclophanes. Royal Society of Chemistry, Cambridge

Eliel EL, Wilen SH, Mander LN (1993) Stereochemistry of organic compounds. Wiley, New York

Garcia-Tellado F, Goswami S, Chang S-K, Geib SJ, Hamilton AD (1990) Molecular recognition – a remarkably simple receptor for the selective complexation of dicarboxylic acids. J Am Chem Soc 112:7393–7394

Gasparrini F, Misiti D, Villani C, Borchart A, Burger, MT, Still WC (1995) Enantioselective recognition by a new chiral stationary phase at receptorial level. J Org Chem 60:4314–4315

Hiyama T (1998) Organosilicon compounds in cross-coupling reactions. In: Diederich F, Stang PJ (eds) Meta-catalyzed cross-coupling reactions. Wiley-VCH, Weinheim, pp 421–453

Jansen JFGA, Meijer EW, de Brabander-van den Berg EMM (1995) The dendritic box: shape-selective liberation of encapsulated guests. J Am Chem Soc 117:4417–4418

Jorgensen WL, Pranata J (1990) Importance of secondary interactions in triply hydrogen bonded complexes: guanine-cytosine vs. uracil-2,6-diaminopyridine. J Am Chem Soc 112:2008–2010

Krogsgaard-Larsen P, Hansen JJ (eds) (1992) Excitatory amino acid receptors: design of agonists and antagonists. Ellis Horwood, New York

Kühn-Velten WN, Kempfle MA (1993) Characterization of the hydrophobic interaction of steroids with endoplasmatic reticulum membranes by quenching of 6,8(14)-bisdehydro-17α-hydroxyprogesterone fluorescence. Biochim Biophys Acta 1145:185–190

Lehn JM (1995) Supramolecular chemistry, concepts, and perspectives, a personal account. VCH, Weinheim

Lemieux RU (1991) Carbohydrate antigens. ACS Symp Ser 519:5–58

Lustenberger P, Martinborough E, Mordasini-Denti T, Diederich F (1998) Geometrical optimization of 1,1'-binaphthyl receptors for enantioselective molecular recognition. J Chem Soc, Perkin Trans 2:747–761

Marti T, Peterson BR, Fürer A, Mordasini-Denti T, Zarske J, Jaun B, Diederich F, Gramlich V (1998) Macrotricyclic steroid receptors by Pd(0)-catalyzed cross-coupling reactions: dissolution of cholesterol in aqueous solution and investigations of the principles governing selective molecular recognition of steroidal substrates. Helv Chim Acta 81:109–144

Martinborough E, Mordasini-Denti T, Castro PP, Wyman TB, Knobler CB, Diederich F (1995) Chiral 1,1'-binaphthyl molecular clefts for the complexation of excitatory amino-acid derivatives. Helv Chim Acta 78:1037–1066

Mattei S, Seiler P, Diederich F, Gramlich V (1995) Dendrophanes: water-soluble dendritic receptors. Helv Chim Acta 78:1904–1912

Mattei S, Wallimann P, Kenda B, Amrein W, Diederich F (1997) Dendrophanes: water-soluble dendritic receptors as models for buried recognition sites in globular proteins. Helv Chim Acta 80:2391–2417

Mitchell TN (1997) Organotin compounds in cross-coupling. In: Diederich F, Stang PJ (eds) Metal-catalyzed cross-coupling reactions. Wiley-VCH, Weinheim, pp 167–202

Mohamadi F, Richards NGJ, Guida WC, Liskamp R, Lipton M, Caufield C, Chang G, Hendrickson T, Still WC (1990) MacroModel – an integrated software system for modeling organic and bioorganic molecules using molecular mechanics. J Comput Chem 11:440–467

Muehldorf AV, Van Engen D, Warner JC, Hamilton AD (1988) Aromatic-aromatic interactions in molecular recognition: a family of artificial receptors for thymine that shows both face-to-face and edge-to-face orientations. J Am Chem Soc 110:6561–6562

Murakami Y, Kikuchi J, Hisaeda Y, Hayashida O (1996) Artificial enzymes. Chem Rev 96:721–758

Neidlein U, Diederich F (1996) Selective complexation of disaccharides by a novel D_2-symmetrical receptor in protic solvent mixtures. Chem Commun 1493–1494

Newkome GR, Moorefield CN, Vögtle F (1996) Dendritic molecules: concepts, syntheses, perspectives. VCH, Weinheim

Obst U, Banner DW, Weber L, Diederich F (1997) Molecular recognition at the thrombin active site: structure-based design and synthesis of potent and selective thrombin inhibitors and the X-ray crystal structures of two thrombin-inhibitor complexes. Chem Biol 4:287–295

Peterson BR, Diederich F (1994) Dissolution of cholesterol in water by a synthetic receptor. Angew Chem Int Ed Engl 33:1625–1628

Peterson BR, Mordasini-Denti T, Diederich F (1995) Cavity depth and width effects on cyclophane-steroid recognition: molecular complexation of cholesterol and progesterone in aqueous solution. Chem Biol 2:139–146

Pieters RJ, Cuntze J, Bonnet M, Diederich F (1997) Enantioselective recognition with C_3-symmetric cage-like receptors in solution and on a stationary phase. J Chem Soc, Perkin Trans 2:1891–1900

Quiocho FA (1989) Protein-carbohydrate interactions: basic molecular features. Pure Appl Chem 61:1293–1306

Sharff AJ, Rodseth LE, Szmelcman S, Hofnung M, Quiocho FA (1995) Refined structures of two insertion/deletion mutants probe function of the maltodextrin binding protein. J Mol Biol 246:8–13

Sharon N, Lis H (1993) Carbohydrates in cell recognition. Sci Am 268(1):74–81

Szejtli J (1988) Cyclodextrin technology. Kluwer, Dordrecht

Vermersch PS, Lemon DD, Tesmer JJG, Quiocho FA (1991) Sugar-binding and crystallographic studies of an arabinose-binding protein mutant (Met[108]Leu) that exhibits enhanced affinity and altered specificity. Biochemistry 30:6861–6866

Vögtle F (1990) Cyclophan-Chemie: Synthesen, Strukturen, Reaktionen: Einführung und Überblick. Teubner, Stuttgart

Watkins JC, Collingridge GL (1989) The NMDA receptor (international symposium). Oxford University Press, Oxford

Wallimann P, Seiler P, Diederich F (1996) Dendrophanes: novel steroid-recognizing dendritic receptors. Helv Chim Acta 79:779–788

Wallimann P, Marti T, Fürer A, Diederich F (1997) Steroids in molecular recognition. Chem Rev 97:1567–1608

Wong CH, Halcomb RL, Ichikawa Y, Kajimoto T (1995) Enzymes in organic synthesis – application to the problems of carbohydrate-recognition. Angew Chem Int Ed Engl 34:412–432 (part 1), 521–546 (part 2)

Zimmerman SC, Murray TJ (1994) Hydrogen bonded complexes with the AA·DD, AA·DDD, and AAA·DD motifs: the role of three centered (bifurcated) hydrogen bonding. Tetrahedron Lett 35:4077–4080

4 Molecular Recognition of DNA by Ecteinascidin 743

B.M. Moore II, F.C. Seaman, and L.H. Hurley

4.1   Introduction 81
4.2   Determination of a NMR-Based Model
        for the Ecteinascidin 743–DNA Adduct 82
4.2.1   Preparation of the Et 743–12-mer DNA Adduct 82
4.2.2   One- and Two-Dimensional NMR Analysis
        of the Et 743–DNA Adduct 83
4.2.3   Molecular Modeling 85
4.3   Mechanism for the Catalytic Activation of Et 743
        and Its Subsequent Alkylation of Guanine N2 88
4.3.1   Carbinolamine-Containing Antibiotics 88
4.3.2   Preparation of the Et 743–12-mer DNA Adduct 89
4.3.3   Determination of the Protenation State
        of the Et 743–DNA Adduct 89
4.3.4   Mechanism for Catalytic Activation of Et 743 91
References 94

4.1 Introduction

The Ecteinascidins (Ets; Rinehart et al. 1990), extremely potent antitumor agents isolated from extracts of the marine tunicate *Ecteinascidia turbinata*, exhibit promising efficacy in several human xenograft models in mice (Sakai et al. 1992, 1996). Their structural novelty prompted researchers to isolate new Et analogs (Sakai et al. 1996), determine the structure (Guan et al. 1993) and absolute configuration of several Ets

1   2

(Sakai et al. 1996), and complete the total synthesis of Et 743 (**1**) (Corey et al. 1996). The first Et to advance to clinical trials is **1** (Sakai et al. 1996); however, the mechanism of antitumor activity remains unclear. Bioassays using purified Ets demonstrated inhibitory activity toward DNA and RNA polymerases (Sakai et al. 1996). Sequence-selective high-affinity binding of **1** to duplex DNA (Pommier et al. 1996) suggests a mechanism of action involving DNA interactions. Additionally, the reactive carbinolamine of **1** is analogous to that found in known guanine N2 (GN2) DNA alkylating agents (Remers and Iyengar 1995). The DNA-reactive saframycins (**2**) are structurally similar to the A and B units of **1**, and based on this similarity theoretical models of **1** bound to DNA have been proposed (Guan et al. 1993).

4.2 Determination of a NMR-Based Model for the Ecteinascidin 743–DNA Adduct

4.2.1 Preparation of the Et 743–12-mer DNA Adduct

Et 743 is reported to react with 5'-GGG, 5'-GGC, and 5'-AGC DNA sequences (Pommier et al. 1996); therefore, in order to assess alkylation selectivity, **1** was reacted with an oligonucleotide containing 5'-GGC (strand 1) and 5'-AGC (strand 2) alkylation sites.

A single oligonucleotide–drug adduct was obtained in which **1** selectively alkylated the 5'-AGC sequence. A 12-mer oligonucleotide, [d(CGTAAGCTTACG)]$_2$, was prepared via phosphoramidite chemistry (Gait 1984) and then reacted with **1** to generate a 1:1 drug–DNA adduct (Moore et al. 1997).

4.2.2 One- and Two-Dimensional NMR Analysis of the Et 743–DNA Adduct

Non-exchangeable proton to proton connectivities in the adduct were determined using two-dimensional (2D) nuclear Overhauser effect spectroscopy (NOESY), correlation spectroscopy (COSY), and total correlation spectroscopy (TOCSY) experiments. Exchangeable protons were studied in an H$_2$O:D$_2$O (9:1) mixture via 2D NOESY experiments. The resulting spectra exhibited well-resolved cross-peaks for both **1** and the oligomer. Total assignment of the 12-mer oligonucleotide cross-peaks was achieved through established methods (Hare et al. 1983), indicating that a single species was present in solution. Intramolecular NOEs for **1** were then used to assign the drug resonances; 47 residual cross-peaks were identified as **1** to DNA intermolecular contacts.

Analysis of **1** to 12-mer NOEs in the non-exchangeable NOESY spectrum yielded several critical connectivities that permitted the positioning of **1** in the minor groove (Fig. 1). The NOEs between 5AH2 and 12NMe plus the 17AH2 to H23 A contacts confirmed the presence of **1** in the minor groove, with the A unit to the 5' side of the alkylated 6G and the B unit to the 3' side. Units A and B are closely associated with the DNA strand opposite the 6GH2 alkylation, showing NOE connectivities between the 18G, 19C, 20T, and 21T protons and the 6Me, 5OAc, H23, H4, and H11 protons. Upfield shifts of 19CH1', 19CH2', 19CH2'', and 18GH1' by 1.07, 0.55, 0.66, and 0.25 ppm, respectively, caused by the aromatic shielding cone over the 19C and 18G deoxyribose, provide additional evidence for the positioning of unit B. Interactions of **1** with the alkylated oligonucleotide strand are evidenced by NOEs between DNA and H21 and H22. The H21 proton shows NOEs into 6GH1', 7CH1', 8TH1', 7CH2'/H2'', 7CH6, and 8TH6. NOEs of H22A and H22B into 7CH1' and 8TH3' are also consistent with **1** being centered around 6G of the oligonucleotide. The surprising observation is the

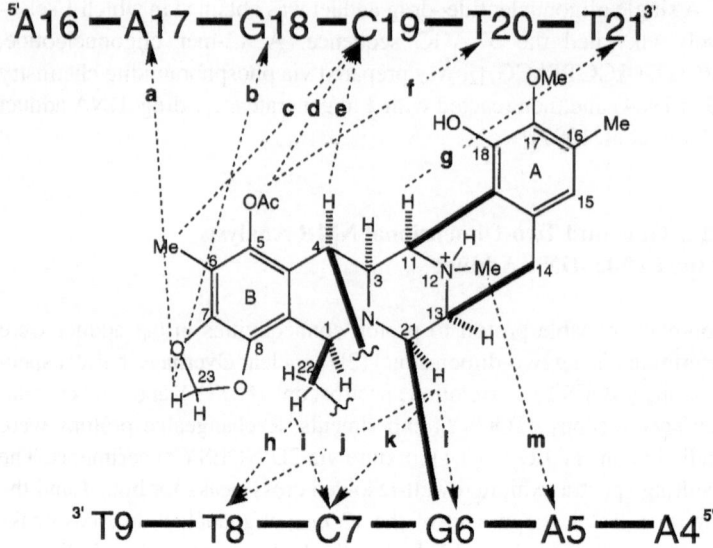

Fig. 1. Schematic model of **1** bound to the 12-mer showing the critical NOE cross-peaks (*arrows*) that define the orientation of **1** in the minor groove of DNA (unit C of **1** has been omitted for clarity). Specific connectivities are: (*a*) 17AH2 to H23 A, (*b*) 18GH1' to H23 A, (*c*) 19CH4'/H5'/H5'' to 6Me, (*d*) 19CH3'/H4' to OAc, (*e*) 19CH2'/H2'' to H4, (*f*) 20TH6 to OAc, (*g*) 20TH1'/H2''/H4' to H11, (*h*) 21TH5' to H11, (*i*) 8TH3' to H22B, (*j*) 7CH1' to H22 A/22B, (*k*) 8TH1' to H21, (*l*) 7CH2'/2'' to H21, (*m*) 6GH1' to H21, and (*n*) 5AH2 to 12NMe

presence of only a single weak oligonucleotide-**1** NOE (8TH3'–7'OMe) involving unit C. The broadness observed in the intramolecular NOEs of unit C suggests that it is interconverting between conformations, i.e., not specifically associating with a site on the DNA. We believe this is evidence that unit C is perpendicularly projected above the minor groove.

The exchangeable proton NOESY spectrum contained the predicted amino and imino interbase connectivities, with an additional resonance at 9.57 ppm. We have observed that alkylation of GN2 by tomaymycin (Boyd et al. 1990) and anthramycin (unpublished results) results in downfield shifts to 8.9 and 9.2 ppm, respectively, of the remaining N2

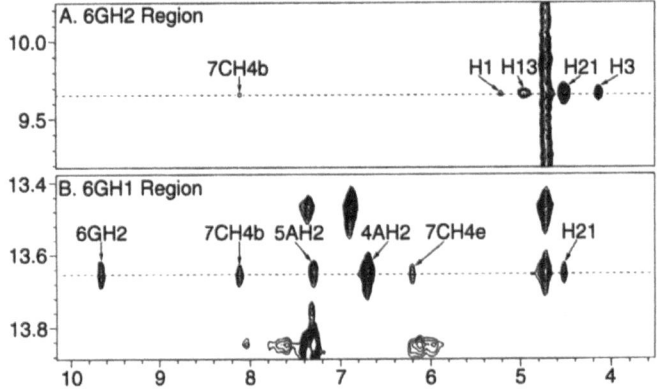

Fig. 2A,B. Connectivities of exchangeable protons in the 1–12-mer adduct.
A Partial 2D water NOESY expansion contour plot of the 6GH2 cross-peaks
into 1 and 7C protons. B Partial 2D water NOESY expansion contour plot of
the 6GH1 cross-peaks into 6GH2 and DNA. The broadness of some of the
cross-peaks is due to the overlap of 21TH3 to 4AH2 and 5AN2

proton; therefore, the 9.57 resonance was assigned to 6GH2. The 6G2
proton showed an NOE to 19CH4 and 6GH1, and the latter showed the
expected intermolecular NOEs to 7CH4b, 7CH4f, and 5AH2 (Fig. 2).
The through-space interactions between 6GH2 and **1** are associated with
H1, H3, H13, and H21 that surround the carbinolamine in **1** (Fig. 2). We
believe that these data, combined with the non-exchangeable NOESY
data, support the proposed alkylation of 6GN2 by **1** and the role of the
carbinolamine in this reaction, and define the position of **1** in the minor
groove of the oligonucleotide.

4.2.3 Molecular Modeling

We generated a model of **1** covalently bound to 6GN2 of the
[d(CGTAAGCTTACG)]$_2$ oligonucleotide using solvated molecular dynamics (Moore et al. 1997; 100 ps, AMBER version 4.1; Pearlman et al.
1995). The resulting model (Fig. 3) is consistent with the NMR data and
generally consistent with the existing theoretical models (Guan et al.

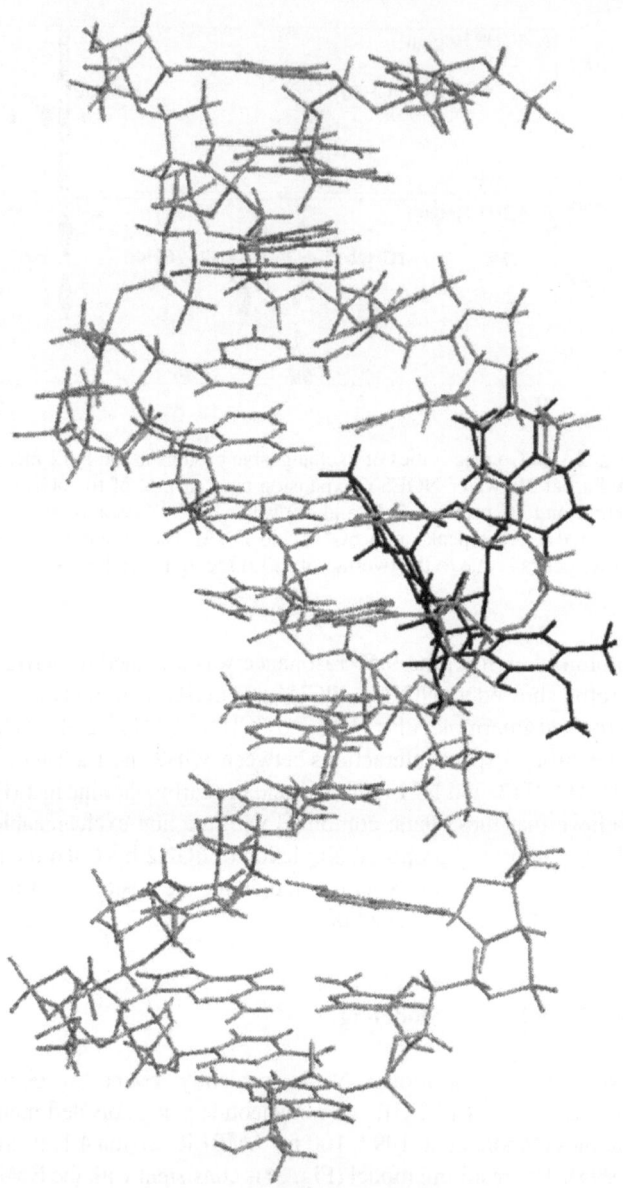

Fig. 3. Stereoview of a molecular model of the 1–oligonucleotide adduct derived from molecular dynamics analysis. The drug is shown in black

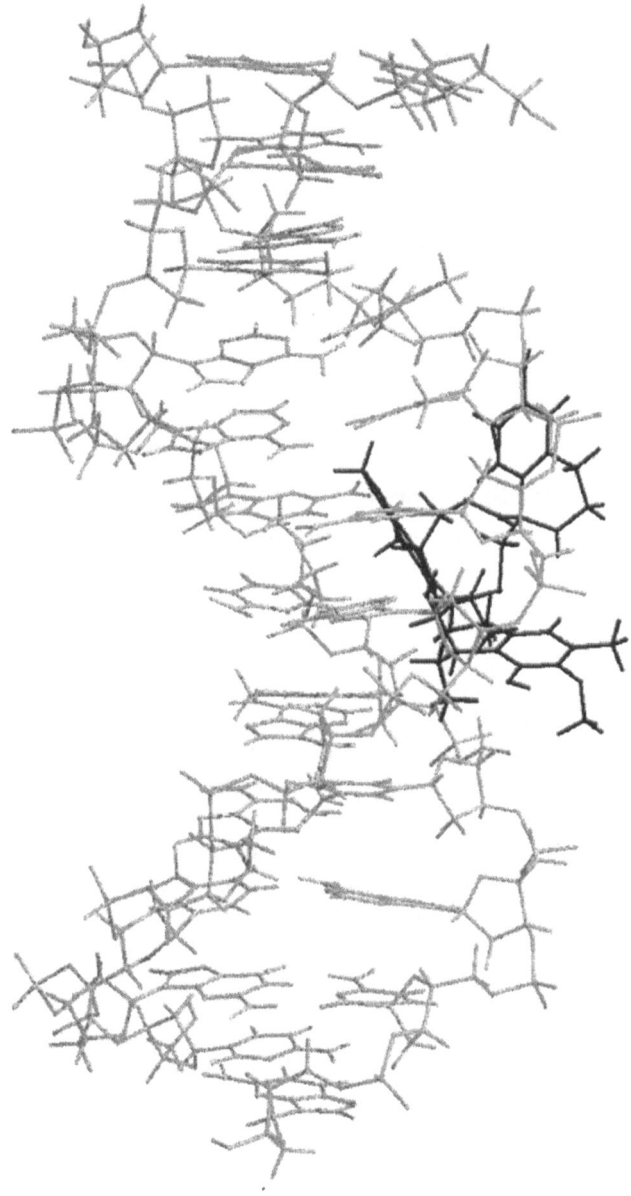

Fig. 3. Legend see p. 86

1993); however, there are differences between the model proposed here
and that previously proposed (Guan et al. 1993). Our model positions
unit B deeper into and closer to the wall of the non-alkylated strand than
previously proposed (Guan et al. 1993) while maintaining the hydrogen
bond from the dioxymethylene oxygen to 18GH2 (Fig. 3). Unit A is
perpendicular to the helical axis and oriented to allow a hydrogen bond
between 18OH and 21TO1' (Fig. 3). The AT base pair to the 5' side of
the alkylation site shifts the hydrogen bonds of the protonated 12NMe to
21TO2 and 21TO1', as compared to the proposed amino/carbonyl pattern in the GC base pairs in the Wang model (Guan et al. 1993). These
ring orientations and hydrogen bonds are, in part, responsible for directing ring C centrally out of the minor groove, thereby restricting interactions with either side of the minor groove.

4.3 Mechanism for the Catalytic Activation of Et 743 and Its Subsequent Alkylation of Guanine N2

4.3.1 Carbinolamine-Containing Antibiotics

The alkylation of GN2 in duplex DNA is well established for a variety
of carbinolamine-containing antibiotics, such as **1** (Pommier et al.
1996), saframycin S (**2**) (Rao and Lown 1992), napthyridinomycin (**3**)
(Zmijewski et al. 1985), and anthramycin (**4**) (Hurley et al. 1977; Lown
and Joshua 1979). The chemical reactivity of these agents has been
proposed to reside in the iminium intermediate generated by the general
acid-mediated dehydration of the carbinolamine moiety (Hurley et al.
1977; Lown and Joshua 1979; Zmijewski et al. 1985). While the mechanism for the formation of the iminium intermediate in bulk solution is a

3   4

general chemical reaction, both the source of general acid in the minor groove of DNA and the protonation state of the adduct remain unidentified. Examination of the carbinolamine antibiotic structures **1–4** shows that basic nitrogens, which should bear a proton at physiological pH, are in close proximity to the hydroxyl group and therefore could serve as a proton source when buried in the minor groove of DNA. Our approach to defining the mechanism of activation for **1** was to determine the protonation state of the drug and DNA in the covalent adduct, then utilize mass and charge balance to propose a mechanism leading to alkylation.

4.3.2 Preparation of the Et 743–12-mer DNA Adduct

Our earlier studies demonstrated that a 1:1 drug–DNA adduct can be prepared by reacting **1** with the 12-mer oligonucleotide [d(CGTAAGCTTACG)]$_2$ (Zimmer and Crothers 1995) containing the preferred AGC alkylation site (Moore et al. 1997). To fully assess the protonation state of the drug and DNA, an isotopically labeled (^{13}C, ^{15}N) 12-mer was synthesized according to the method of Crothers (Zimmer and Crothers 1995), and the natural abundance duplex was prepared via phosphoramidite chemistry (Gait 1984). Both the labeled and natural abundance DNAs and the **1**–DNA adducts were investigated by high-field NMR spectroscopy.

4.3.3 Determination of the Protenation State of the Et 743–DNA Adduct

Nitrogen atoms N2 and N12 of **1** and 6GN2 were the most likely candidates for protonation in the **1**–DNA adduct since they are centered around C21 of **1**, the site of nucleophilic attack of 6GN2. To assess the protonation of **1** in the adduct, the natural abundance **1**–DNA adduct was studied utilizing COSY, TOCSY, and NOESY experiments in an H$_2$O–D$_2$O (9:1) solvent system. The TOCSY (Fig. 4) and COSY of **1**–DNA have a single cross-peak in the NH chemical shift range (8.05 ppm), showing a connection to 12NMe; additionally, this peak was absent in the D$_2$O spectrum, indicating that the proton was ex-

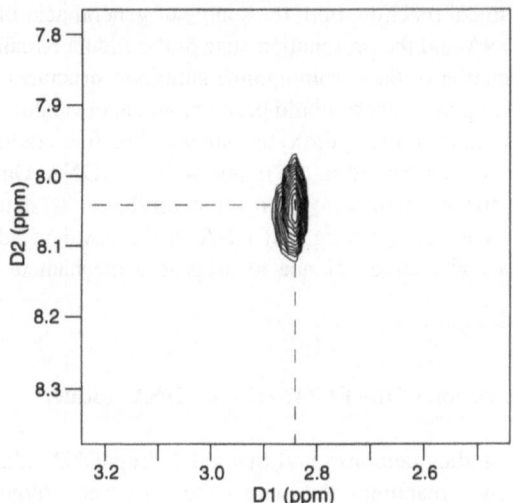

Fig. 4. Partial 2D water TOCSY expansion contour plot of the cross-peak of 12NH into the 12NMe in the 1–DNA adduct

changeable. These data showed that 12N of **1** is protonated in the covalent adduct, and the relative intensity of the cross-peaks suggested that the proton was in slow exchange, i.e., probably hydrogen-bonded. The absence of cross-peaks in the NH region to H1, H3, and H21 of **1** in the TOCSY and COSY spectra indicates that N2 is not protonated in the adduct. This observation was consistent with the reported steric crowding around N2 in the crystal structure of Et 743 N^{12}-oxide (Guan et al. 1993) and by the fact that a second protonation in a piperazine-like system is unlikely above pH 2.0 (Delpuech and Martinet 1971).

Heteronuclear multiple quantum correlation (HMQC) and heteronuclear multiple bond correlation (HMBC) NMR spectroscopy in an H_2O–D_2O (9:1) solvent system was utilized to assess the protonation state of the DNA in the **1**–DNA adduct. Figure 5 shows the $^{15}N^1H$ chemical shift changes for 6GN2H associated with covalent bonding of **1** to DNA. Prior to drug modification, the NH cross-peak of 6G is in the usual range (proton 6.6–9.0 ppm, nitrogen 75 ppm) for duplex DNA (Zimmer and Crothers 1995); however, covalent modification of 6GN2 by **1** shifts the 6GN2H 1.95 ppm downfield, while 6GN2 is only shifted

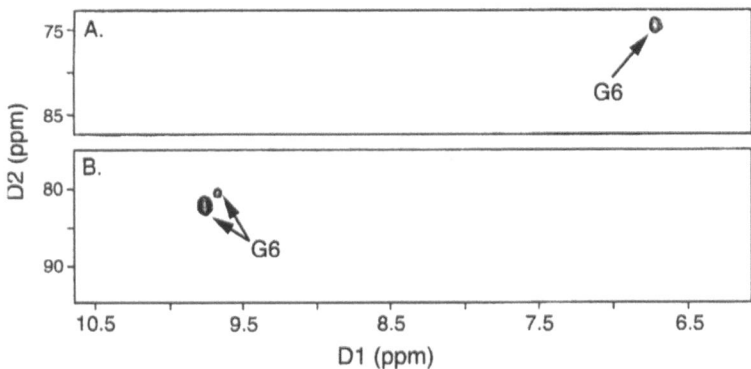

Fig. 5A,B. Proton nitrogen correlations for the isotopically labeled DNA and the 1–DNA adduct. **A** Partial HMQC expansion contour plot showing the 6GN2-H correlations of the duplex DNA. **B** Partial HMQC expansion contour plot showing the 6GN2-H correlations of the 1–DNA adduct

6.12 ppm. The presence of a minor cross-peak in the 6GNH region most likely reflects the conformational flexibility of the C ring of **1** in the covalent adduct (manuscript in preparation). The modest downfield shift in 6GN2, compared with a 40–60 ppm upfield shift expected if 6GN2 were protonated (Martin et al. 1981), clearly demonstrated that 6GN2 is not protonated. The lack of DNA protonation was evidenced by the absence of 40–60 ppm chemical shift changes in the ^{15}N spectra (data not shown) and the absence of thermal strand breakage (Guan et al. 1993). The downfield shift in the hydrogen dimension therefore results from elimination of the time-averaged chemical shift of the 6GNH$_e$ and 6GNH$_b$ to a cross-peak arising from just the 6GNH$_b$ in the covalent adduct.

4.3.4 Mechanism for Catalytic Activation of Et 743

Scheme 1 shows a mechanism for the activation of **1** and covalent modification of 6GN that is consistent with the NMR data. The approach of **1** into the minor groove would be accompanied by desolvation of **1**, which in turn would increase the strength of the hydrogen bond

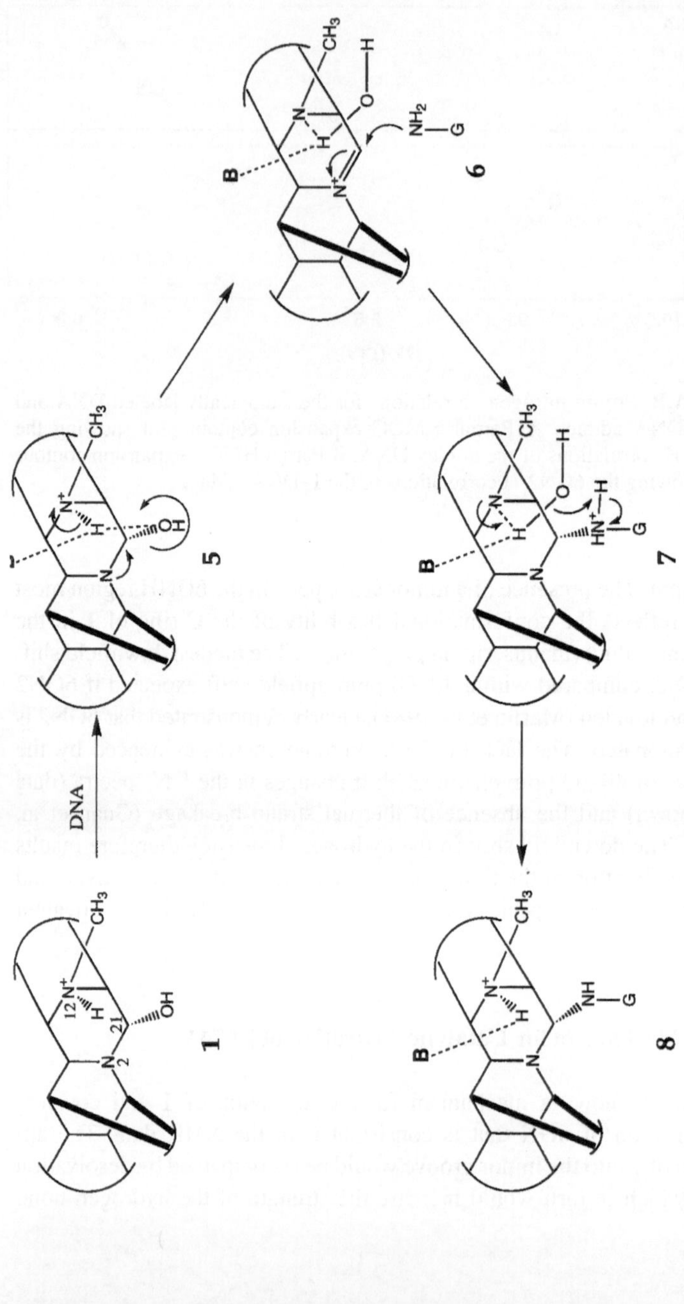

Scheme 1. Mechanism for the catalytic activation of **1** and alkylation of 6GN2. *Dotted lines* represent hydrogen-bond interactions, and *B* represents a DNA base accepting a hydrogen bond

between the protonated 12NH and 21OH. The close proximity (2.54 Å) of 12NH to 21OH and the partial sp^2 character of 2N in **1** (Guan et al. 1993) should be sufficient to catalyze the dehydration of the carbinolamine (**5**). Since 12NH participates in hydrogen bonds in the covalent adduct and is believed to play a role in sequence recognition of DNA (F. C. Seaman, unpublished results; Sakai et al. 1992), it is probable that the water remains hydrogen-bonded at this point in the mechanism (**6**). Nucleophilic attack at C21 of the iminium intermediate by 6GN2 results in a charged 6GN (**7**) having a proton with a pK_a expected to be below zero (Shapiro and Gordon 1964). This proton could be transferred either to N2 or N12 of **1** or to a water molecule in the proximity of the protonated 6GN2. Considering the low pK_a of this proton, it could readily transfer to the transiently bound water molecule, setting up a proton shuttle by which the original protonation state of 12N of **1** would be restored and the water molecule expelled as the transition state collapses to the final covalent adduct (**8**). The final structure of the **1**–DNA adduct is consistent with the NMR data and an earlier theoretical activation mechanism of **4** (Hurley et al. 1977; Lown and Joshua 1979); furthermore, the essential hydrogen bonds for DNA recognition and binding are preserved throughout the mechanism.

In this paper, we have proposed a chemical mechanism that gives rise to the alkylation of 6GN2 by **1** in duplex DNA. These data indicate that in the development of ecteinascidin analogs, a proton source should be retained in close proximity to the carbinolamine to catalyze the dehydration of this moiety. This result implies that Nature has developed a mechanism to assure the reactivity of the carbinolamine antibiotics **1–4** by the inclusion of an internal catalytic proton adjacent to the leaving hydroxyl group. Finally, we believe these results not only impact on the development of ecteinascidins but also aid in the de novo design of new DNA-reactive reagents.

Acknowledgement. We thank PharmaMar USA for a sample of Et 743. We are grateful to Steve Sorey for NMR technical assistance. Research was supported by a grant from the National Institutes of Health (CA-49751). We also thank David Bishop for preparing, proofreading, and editing the manuscript.

References

Boyd FL, Stewart D, Remers WA, Barkley MD, Hurley LH (1990) Characterization of a unique tomaymycin-d(CICGAATTCICG)$_2$ adduct containing two drug molecules per duplex by NMR, fluorescence, and molecular modeling studies. Biochemistry 29:2387–2403

Corey EJ, Gin DY, Kania RS (1996) Enantioselective total synthesis of ecteinascidin 743. J Am Chem Soc 118:9202–9203

Delpuech JJ, Martinet MY (1971) Transferts protonique de sels d'ammonium substitues. IV. Vitesses d'inversion sur l'azote par protonation competitive et basicite de N,N'-dimethylpiperazines. Tetrahedron 27:2499–2515

Gait MJ (ed) (1984) Oligonucleotide synthesis – a practical approach. IRL Press, Oxford

Guan Y, Sakai R, Rinehart KL, Wang AH-J (1993) Molecular and crystal structures of ecteinascidins: potent antitumor compounds from the Caribbean tunicate Ecteinascidia turbinata. J Biomol Struct Dyn 10:793–818

Hare DR, Wemmer DE, Chou SH, Drobny G, Reid BR (1983) Assignment of the non-exchangeable proton resonances of d(C-G-C-G-A-A-T-T-C-G-C-G) using two-dimensional nuclear magnetic resonance methods. J Mol Biol 171:319–336

Hurley LH, Gairola C, Zmijewski M (1977) Pyrrolo(1,4)benzodiazepine antitumor antibiotics in vitro interaction of anthramycin, sibiromycin and tomaymycin with DNA using specifically radiolabelled molecules. Biochim Biophys Acta 475:521–535

Lown JW, Joshua AV (1979) Molecular mechanism of binding of pyrrolo(1,4)benzodiazepine antitumour agents to deoxyribonucleic acid – anthramycin and tomaymycin. Biochem Pharm 28:2017–2026

Martin GJ, Martin ML, Gouesnard JP (1981) ^{15}N-NMR spectroscopy. Springer, Berlin Heidelberg New York

Moore BM, II, Seaman FC, Hurley LH (1997) NMR-based model of an ecteinascidin 743–DNA adduct. J Am Chem Soc 119:5475–5476

Pearlman DA, Case DA, Caldwell JC, Wilson RS, Cheatman TE, III, Ferguson GL, Seibel GL, Singh UC, Weiner D, Kollman PA (1995) AMBER 4.1. University of California, San Francisco

Pommier Y, Kohlhagen G, Bailly C, Waring M, Mazumder A, Kohn KW (1996) DNA sequence- and structure-selective alkylation of guanine N2 in the DNA minor groove by ecteinascidin 743, a potent antitumor compound from the Caribbean tunicate Ecteinascidia turbinata. Biochemistry 35:13303–13309

Rao KE, Lown JW (1992) DNA sequence selectivities in the covalent bonding of antibiotic saframycins Mx1, Mx3, A, and S deduced from MPE·Fe(II)

footprinting and exonuclease III stop assays. Biochemistry 31:12076–12082

Remers WA, Iyengar BS (1995) Antitumor antibiotics. In: Foye WO (ed) Cancer chemotherapeutic agents. American Chemical Society, Washington DC, pp 577–679 and references therein

Rinehart KL, Holt TG, Fregeau NL, Stroh JG, Keifer PA, Sun F, Li LH, Martin DG (1990) Ecteinascidins 729, 743, 745, 759 A, 759B, and 770: potent antitumor agents from the Caribbean tunicate Ecteinascidia turbinata. J Org Chem 55:4512–4515

Sakai R, Rinehart KL, Guan Y, Wang AH-J (1992) Additional antitumor ecteinascidins from a Caribbean tunicate: crystal structures and activities in vivo. Proc Natl Acad Sci USA 89:11456–11460

Sakai R, Jares-Erijman EA, Manzanares I, Elipe MVS, Rinehart KL (1996) Ecteinascidins: putative biosynthetic precursors and absolute stereochemistry. J Am Chem Soc 118:9017–9023

Shapiro R, Gordon CN (1964) On the structure of neoguanosine. Biochem Biophy Res Commun 17:160–164

Zimmer DP, Crothers DM (1995) NMR of enzymatically synthesized uniformly 13C15N-labeled DNA oligonucleotides. Proc Natl Acad Sci USA 92:3091–3095

Zmijewski MJ, Miller-Hatch K, Mikolajczak M (1985) The in vitro interaction of naphthyridinomycin with deoxyribonucleic acids. Chem Biol Interact 52:361–375

5 New Tools for Drug Design Based on Protein Ligand Recognition Principles

G. Klebe

5.1   Molecular Recognition and Protein-Ligand Complexes   97
5.2   Structural Knowledge about the Ligand-Protein Interface   99
5.3   Binding Motifs Between Ligands and Building Units of Proteins   99
5.4   Preferred Interaction Sites Between Ligands and Proteins   102
5.5   Distribution of Non-bonded Contacts
         from the Cambridge Structural Database   103
5.6   Computational Methods Exploiting Composite
         Crystal-Field Environments   106
5.7   Conformational Preferences
         Observed in the Cambridge Structural Database   108
5.8   Computational Methods Exploiting Conformational Preferences   112
5.9   Hydrogen-Bond Acceptors in Competitive Situations:
         Which Is the Better Acceptor?   113
5.10   Conclusion   115
References   116

5.1 Molecular Recognition and Protein-Ligand Complexes

Jean-Marie Lehn defined molecular recognition in 1973 as a process involving the binding and selection of substrate(s) by a given receptor molecule linked to a specific function (Lehn 1973). Molecular recognition is therefore binding that serves a particular purpose; it implies a structurally well-defined pattern of intermolecular interactions, requires appropriate shape complementarity, and is accordingly a question of information storage and read-out at the supramolecular level (Lehn 1988).

These definitions can be directly applied to protein-ligand complexes. Enzymes and receptors recognize their ligands such as substrates, agonists, antagonists or allosteric effectors via specific intermolecular interactions, sometimes with extreme selectivity. Through binding and/or chemical transformation, information is transmitted in a biological system. Due to their functional role in biochemical pathways or signal transduction chains, enzymes and receptors display important targets in the treatment of diseases. Modulation of the information flow in such systems might result in a therapeutic principle. In such cases, the drug design process attempts to use the principles of molecular recognition to develop selective inhibitors, synthetic agonists, or antagonists that exploit this structurally well-defined pattern of intermolecular interactions between a ligand and a protein.

In 1991, Jack Dunitz dubbed a crystal a "supramolecule par excellence," held together by the kind of non-covalent binding interactions that are responsible for molecular recognition and complex formation at all levels (Dunitz 1991). Since long-range periodicity in a crystal is a product of directional specific short-range interactions, a crystal is an ordered supramolecule. Over recent years we have witnessed an exponential increase in solved crystal structures, both of proteins and small molecules. This vast and rapidly growing 3D structure database of protein-ligand complexes and small organic molecules contains important information about the stereochemistry of supramolecular assemblies.

In drug design, one attempts to predict the binding (recognition) properties of molecules. Accordingly, this discipline is greatly concerned with understanding the principles of molecular recognition. In 1995, Gautam Desiraju coined the term "supramolecular synthons" as being structural units within supramolecules (Desiraju 1995). They display spatial arrangements of molecular building blocks exhibiting functional groups that generate well-defined patterns of intermolecular interactions. Drug design, typically using computational methods, tries to assemble appropriate supramolecular synthons into a molecular framework that serves as a ligand for a protein. Can we exploit the principles of molecular recognition for the development of new computational tools in drug design? Problems of interest are the prediction of putative ligands in de novo design and docking, their conformational properties, or their similarity in producing comparable biological effects.

5.2 Structural Knowledge about the Ligand-Protein Interface

Structure-based approaches toward the design of new drugs try to evaluate our current knowledge about structural and energetic aspects of protein-ligand interactions. They correlate the existing empirical knowledge in order to derive rules or general principles. They obtain their significance through a systematic evaluation of these data from a statistical point of view. Accordingly, efficient access and reliable retrieval software are required.

An integrated database has only recently been assembled (Hemm et al. 1995), which allows evaluation of ligand-protein complexes within the Protein Data Bank (PDB, Bernstein et al. 1977). The ReliBase tool allows for a simultaneous search of, e.g., protein sequence data along with connectivity data of ligands. Structural properties of bound ligands can be correlated with the residue constitution or geometrical features found at the binding site of a protein.

5.3 Binding Motifs Between Ligands and Building Units of Proteins

The structural organization of aromatic moieties is dominated by the formation of T-shaped, edge-to-face arrays (synthon of aromatic rings) in which C-H bond dipoles at the rim of one ring orient toward the negatively charged carbons of a neighboring ring. These aromatic interactions determine the packing in crystalline benzene. A striking similarity between the non-covalent contact patterns in benzene crystals and the arrangement of aromatic rings in protein-ligand or aromatic host-guest complexes can be found (Klebe and Diederich 1993; Böhm and Klebe 1996). Subsets from the packing in crystalline benzene superimpose nicely with the orientation of aromatic groups at the binding site of protein-ligand complexes (Fig. 1). This information may be useful for predicting putative binding sites for aromatic groups in ligands.

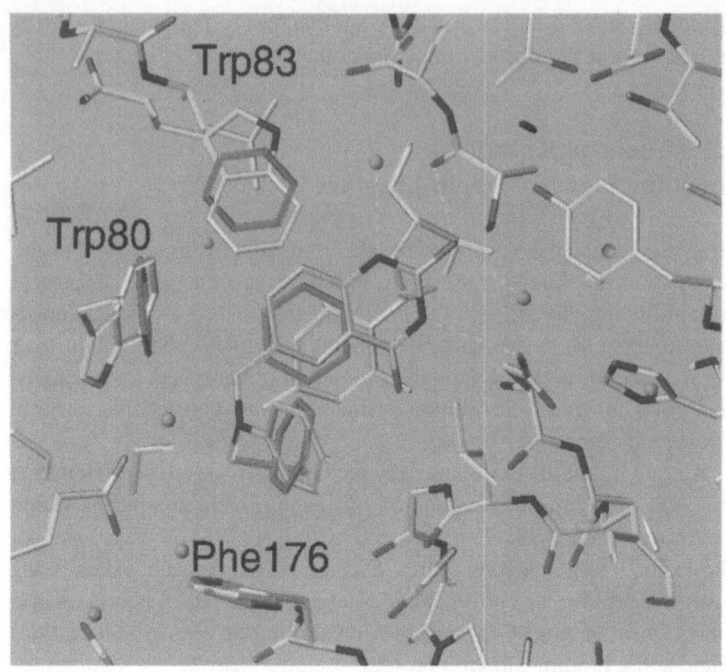

Fig. 1. A subset from the crystal packing in crystalline benzene (gray six-membered rings) has been superimposed on the binding geometry of CB3717, a ligand binding to thymidylate synthase. Two benzene rings fall upon two aromatic groups of the ligand, and three additional ones superimpose with the aromatic portions in Trp 80, Trp 83 and Phe 176

Another preferred recognition site for aromatic moieties appears to occur close to amide bonds (Klebe 1994a). Amide groups are potent hydrogen-bond forming partners within the plane of the amide bond. A variety of structures can be found where the N-H bond dipole is oriented along the normal on the plane through the phenyl ring. However, perpendicular to the amide plane, the amide bond shows mainly hydrophobic properties. Accordingly, a slit-type groove (e.g., the opening between two parallel β-sheets) can accommodate aromatic groups of ligands (Fig. 2). In some examples, one of the flanking amide groups is replaced by a cluster of neighboring aromatic moieties showing the above-mentioned edge-to-face arrangement.

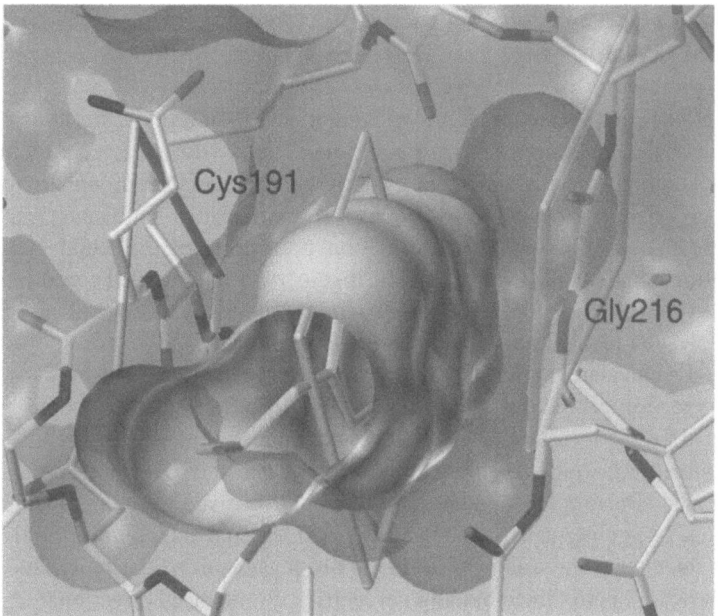

Fig. 2. Two β-strands, oriented parallel to each other (indicated by best planes through the peptide bonds of Cys 191 and Gly 216), provide a slit-type groove that serves as a recognition pocket for a phenyl group of a ligand (*center*, surrounded by the solid solvent accessible surface). This pattern can be observed in several protein-ligand complexes, e.g., in the present example in trypsin with the bound inhibitor *p*-amidino-phenyl pyruvate

Ligands often comprise charged groups. Due to their large polarization, aromatic side chains can favorably interact with positively charged ligand groups, e.g., quaternary ammonium groups. This type of interaction is, for example, found in the complex of acetylcholinesterase with acetylcholine (Sussman et al. 1991). Furthermore, the position at the apex of an α-helix is a favorable binding site for charged groups. Proteins exploit this principle to increase their spatial stability. Often charged amino acid residues are observed at the ends of helices. Similarly, this terminal position can also serve as a favorable binding site for charged groups in ligands (Klebe 1994a). For example, the diphosphate moiety in co-factors like NADPH is located in many enzymes at this position (Hol and Wierenga 1984).

5.4 Preferred Interaction Sites Between Ligands and Proteins

Rational drug design attempts to predict new molecules or portions of molecules that can form favorable interactions with a target protein. Accordingly, putative interaction geometries of ligand functional groups with active site amino acid residues ("supramolecular synthons") must be proposed. Based on this information, predictions on the actual orientation of a ligand at the binding site together with its adopted conformation must be provided. To what extent can database knowledge about molecular recognition assist in these steps of the design process?

To elucidate conformational preferences or patterns and motifs of interaction geometries, data from protein-ligand complexes should be used for the statistical analysis. These complexes are the most relevant source for learning about the protein-ligand interface. However, protein crystal structure analyses are not generally resolved to atomic resolution. As a consequence, the remaining positional errors and uncertainties in the torsion angle adjustment to the residual electron density will only allow for approximate binding geometries. Pronounced directional features in interaction patterns or definite torsional preferences may become smeared out and obscure any detailed conclusions. Furthermore, the protein-ligand data presently available are still limited and therefore do not allow a significant statistical evaluation. On the other hand, an enormous amount of information about the intra- and intermolecular geometry of organic molecules is available in small molecule crystal structures. Generally, the latter structures are resolved to atomic resolution with low positional errors. They have been compiled in the Cambridge Structural Database (CSD) and are easily handled using computational techniques (Allen et al. 1983). Sophisticated retrieval software has been developed that allows the extraction of systematic features or motifs in orientational and conformational preferences around particular functional groups or within torsional fragments.

5.5 Distribution of Non-bonded Contacts
from the Cambridge Structural Database

The connectivity of a particular probe fragment under consideration, e.g., a phenol group (as mimic for a tyrosine), is coded for retrieval. Appropriate interacting groups such as hydrogen donor (NH, OH groups), acceptor groups (O, N), or non-polar groups (aromatic C-H, aliphatic C-H) are defined and searched for in contact distance (Klebe 1994b). The coordinates of the selected "hit" fragments are stored. A common orientation is achieved by a least squares superposition of the atoms of the probe fragment that are similar in all structures. The coordinates of all contact atoms are merged into one combined assembly. The composite picture obtained represents the entire crystal-field environment and can serve as a reliable guide toward the mapping of putative recognition sites around ligands (Fig. 3).

In order to demonstrate the relevance of the small molecule data for the amino acid/ligand functional group recognition in protein-ligand complexes, the corresponding crystal-field environments were superimposed onto the amino acid residues that orient their functional groups toward the accessible surface of the active site (Klebe 1994b). Any sites from these distributions falling next to atoms of the protein were rejected from the analysis. The remaining contact atoms map-out putative interaction sites in the binding pocket. A comparison with crystallographically resolved protein-ligand complexes shows that the observed positions of ligand atoms interacting with the protein coincide convincingly with regions frequently occupied by bonding partners in organic crystal structures (Fig. 4).

In cases where no structural information about a particular target protein is available, ligands binding to the same receptor can be brought into a common orientation by geometrically superimposing similar functional groups (pharmacophore). Mapping composite crystal-field environments onto these functional groups in various ligands allows one to characterize putative areas in space where these ligands might interact with the same functional groups of a common protein receptor (Klebe 1994b). This approach can be used for the superposition of structurally deviating ligands and provides criteria for the comparison of putative interaction properties. Subsequently, it can be used to derive hypothetical receptor models.

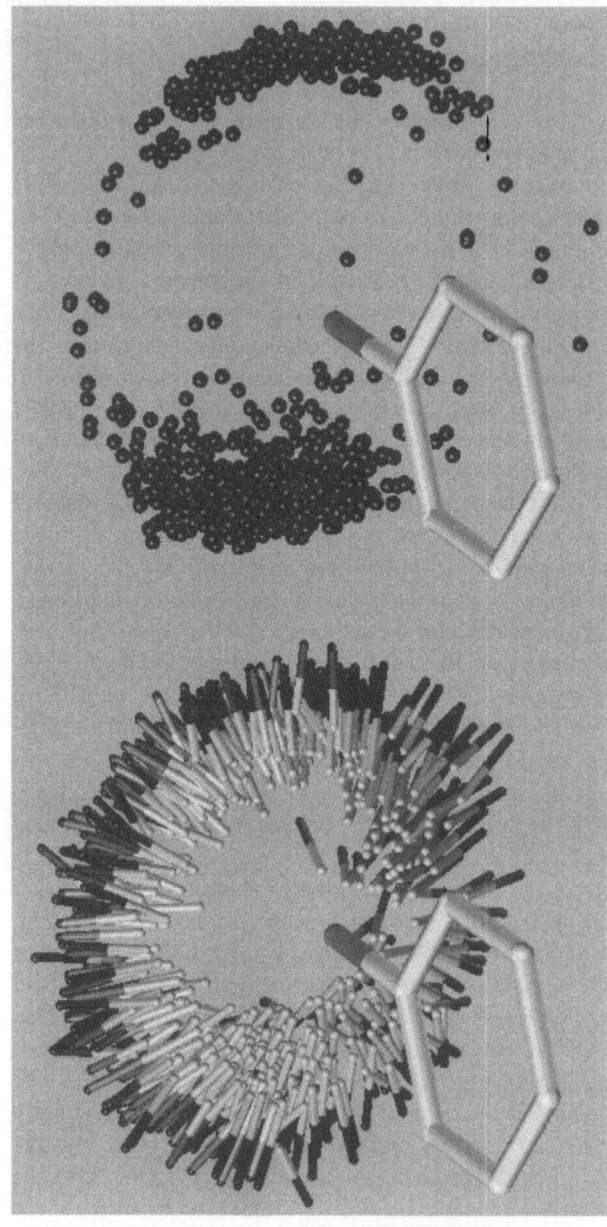

Fig. 3. Composite picture of the contact geometries around a phenolic OH group to map out possible spatial orientations of interactions with hydrogen-bond donors (*left*) or acceptors (*right*). Data were retrieved from small molecule crystal structures. The coordinates of the probe functional group together with the contacting atoms [NH, OH (O, N *black*, H *gray*); N, O (*black balls*)] were extracted. A common orientation of all considered examples was achieved by a least squares superposition of the atoms of the phenyl rings. The hydrogens directly bound to the phenolic framework have been omitted for clarity

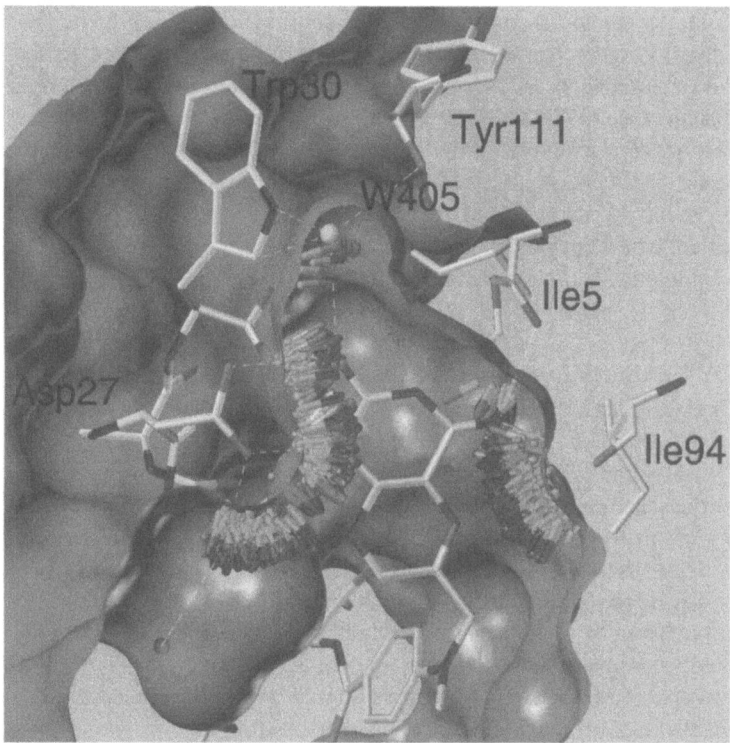

Fig. 4. Section through the binding pocket of dihydrofolate reductase (*E. coli*) with the bound ligand methotrexate: view perpendicular to the pteridine moiety, hydrogen bonds as *dashed lines*. Water 405 mediates an H-bond from the protein to the ligands. Matched onto the amino acid residues oriented toward the binding site are distributions of corresponding composite crystal-field environments of non-bonded interactions. After rejecting contact atoms which penetrate into atoms of the protein, the indicated positions remain as putative binding sites about the carboxylate group of Asp 27, the backbone carbonyl of Ile 5, Ile 94, and Tyr 111, the side-chain oxygen of Thr 113 (NH, OH bond vectors, H *gray*, O, N *black*) and the indole NH of Trp 30 (N, O atom positions, *black balls*)

In a contemporary project at the Cambridge Crystallographic Data Center, a comprehensive library of several hundred composite crystal-field environments around various functional groups has been compiled (Bruno et al. 1997). This collection of different contact geometries has been made available with the IsoStar tool. This database contains the actual distributions, together with associated density representations. Each entry is hyperlinked to the Cambridge Structural Database, a feature which enables immediate access to the original crystal data that went into the data compilation.

5.6 Computational Methods Exploiting Composite Crystal-Field Environments

The information contained in composite crystal-field environments can be translated into rules that serve as guidelines for the automatic docking of small molecules or molecular fragments into the active site of proteins (Klebe 1994b; Böhm and Klebe 1996).

The de novo design tool LUDI (Böhm 1992a,b) heavily exploits this information. In a first step, the program calculates putative interaction sites according to their suitability in forming hydrogen bonds or filling hydrophobic pockets. These interaction sites are generated as sets of equally spaced points or vectors in those spatial regions found to be favorable according to the corresponding crystal-field environments. Next, LUDI fits molecular fragments from a database onto these sites. Subsequently, it can also be used to connect additional fragments to already placed molecular portions. Alternatively, two independently fitted fragments can be connected by a suitable bridge linker. Finally, the structures obtained are scored using an empirically derived scoring function to predict binding affinities (Böhm 1994, 1998; Böhm and Klebe 1996). Since the program does not perform any elaborate geometry optimizations or force-field calculations, it is very fast and tolerant to small uncertainties in the experimentally determined protein structures.

The program FlexX, a tool for flexible docking, also makes use of rules derived from composite crystal-field environments (Rarey et al. 1996). To describe the structural requirements for docking, a particular interaction between two complementary groups in a ligand and a protein is defined in the following way: each functional group in the ligand and

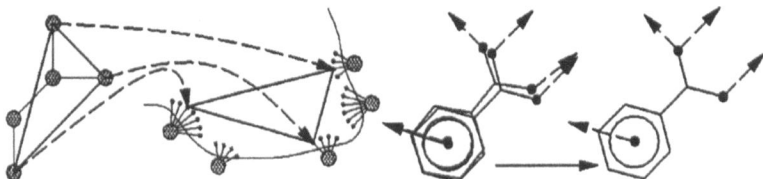

Fig. 5. Geometrically equivalent triangles of interaction centers are searched for around the base fragment and in the binding site *(left)*. For each satisfactory matching pair a transformation is calculated and similar solutions are clustered and subsequently merged *(right)*

in the active site is assigned to a particular type of interaction geometry. These geometries are classified according to the spatial distribution observed in corresponding composite crystal-field environments. As in the LUDI program, they are expressed in terms of sets of discrete interaction sites. In the docking algorithm, a protein-ligand interaction is considered possible if corresponding sites around the two interacting functional groups fall next to each other. In the following step, a base fragment is placed into the binding site using a pose-clustering algorithm. Triplets of interaction sites are formed between the receptor and the ligand (Fig. 5). From each triplet, a transformation is computed such that all three interactions between protein and ligand can be satisfied simultaneously. Since several closely similar solutions can occur, these solutions are clustered. In the next step, an incremental build-up procedure is applied in order to reconstruct the entire ligand from individual fragments. Conformational flexibility is considered, simultaneously optimizing putative interactions with the remaining interaction sites. At each step, a ranking of the achieved placements is performed. Multiple solutions are tested during the incremental build-up following a tree-type search procedure.

Information from composite crystal-field environments has also been incorporated into an approach to superimpose structurally diverse molecules known to bind to a common protein binding site. This technique is applied in particular when no structural information about the receptor protein is available. To quantify the spatial similarity of two molecules, their shape is approximated by a set of Gaussian-type functions centered at the atomic positions (Kearsley and Smith 1990; Klebe et al. 1994).

Subsequently, for each molecule these functions are associated with a vector of physicochemical properties derived from atom-based descriptors. To consider the hydrogen-bonding facilities of the different molecules, putative donor and acceptor groups are generated around the functional groups of a ligand. They are placed at positions where H-bonding partners would be expected according to the corresponding composite crystal-field environments. To consider the spread in these distributions, similar to the shape description, Gaussian functions are located at the centers of these distributions. Finally, the similarity of the two molecules is computed by evaluating the scalar product of the above-mentioned associated vectors weighted by the overlap of the corresponding Gaussian functions. In an iterative process, the minima of this condition are searched. Alternatively to the six-dimensional optimization (three rotational and three translational degress of freedom) in real space, an optimization in Fourier space has been attempted (Nissink et al. 1997). The major advantage of this approach is the separation of the rotational and translational optimization into two successive three-dimensional searches, substantially speeding up the optimization procedure.

5.7 Conformational Preferences
Observed in the Cambridge Structural Database

Flexible molecules often exhibit several conformations of nearly equal energy. The conformation adopted in any given case will depend on the particular interactions with the surrounding environment and will alter the recognition properties of a molecule (Klebe 1994a). A variety of computational search techniques have been developed to generate a large ensemble of multiple conformers for drug-size molecules (Klebe 1995). They generate low-energy conformers that correspond to local minima on a particular energy surface. The efficiency and reliability of these methods is usually assessed by criteria such as: is the global minimum found, or how many local minima are detected within a particular energy range above the global minimum? In general, the ligand is considered alone, not just for the sake of simplicity but often because no or only limited information is available about the protein binding site. The important question arises as to whether ligands adopt conformations at the binding pocket that correspond to local minima observed in the isolated state. To what extent does a local environment modify conformational properties, especially with respect to the location and relative energy level of local minima on the energy surface? Is it justified to concentrate on local minima and how could they be detected?

In the following, these questions will be addressed empirically. The best approach would be a detailed analysis of experimental protein-ligand complexes. However, as mentioned above, we must recall the limited resolution and accuracy of these structure determinations. The danger exists that they will prevent significant conclusions. An alternative could be to concentrate on the conformation of small molecules adopted in their crystal structures. A comparative analysis of ligand conformations observed in crystalline protein-ligand complexes and small-molecule crystal structures has shown that the reliance on just the crystal structure of a ligand itself is not, in most cases, an infallible and reliable indicator of a binding site geometry (Ricketts et al. 1993; Klebe 1994a; Nicklaus et al. 1995). For rather non-polar and rigid ligands, where the environment is composed by many isotropically distributed atom-atom potentials, a satisfactory correspondence may be given. In these cases, the active site conformation is mainly determined by intramolecular forces and no significant perturbations arise from strong directional interactions with the crystal or active-site environment.

For polar ligands embedded into a network of strongly directional hydrogen bonds, the environment clearly influences the adopted conformation. Accordingly in these cases, the analysis has not been focused on entire molecules but on the study of torsional fragments within molecules. Through the statistical analysis of conformational preferences in these torsional fragments, it was found that data from small-molecule crystal structures provide important information about conformations also adopted at the protein binding site. Instead of using individual structures to suggest a single possible conformation, we rather analyze them in terms of conformational preferences adopted in various molecular environments (Klebe 1994a). Exhaustive searches of the Cambridge Crystallographic Database revealed a library of more than 1000 torsional fragments (Klebe et al. 1998). This library provides a convincing ability to generate relevant binding-site geometries. They appear to be

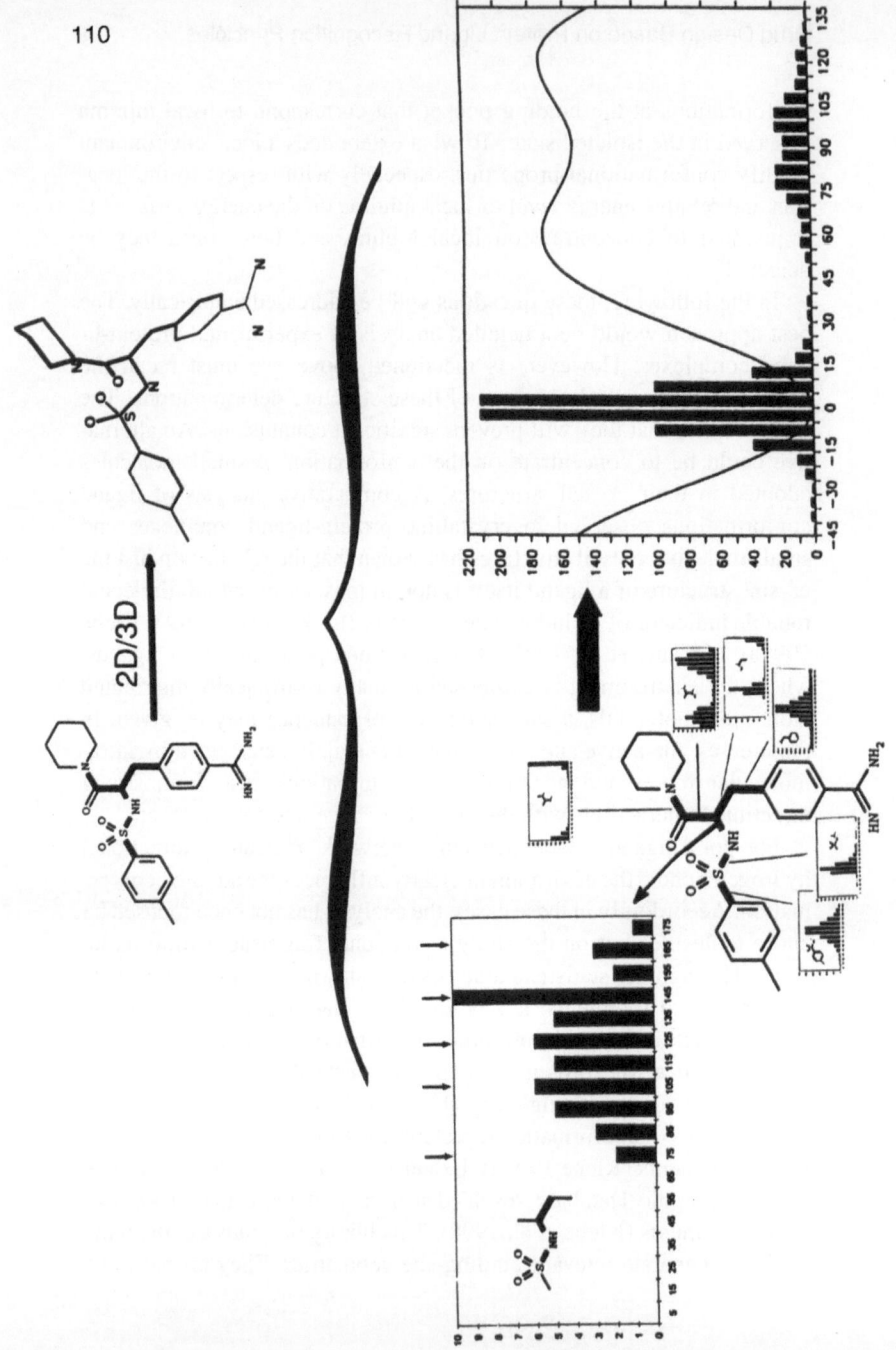

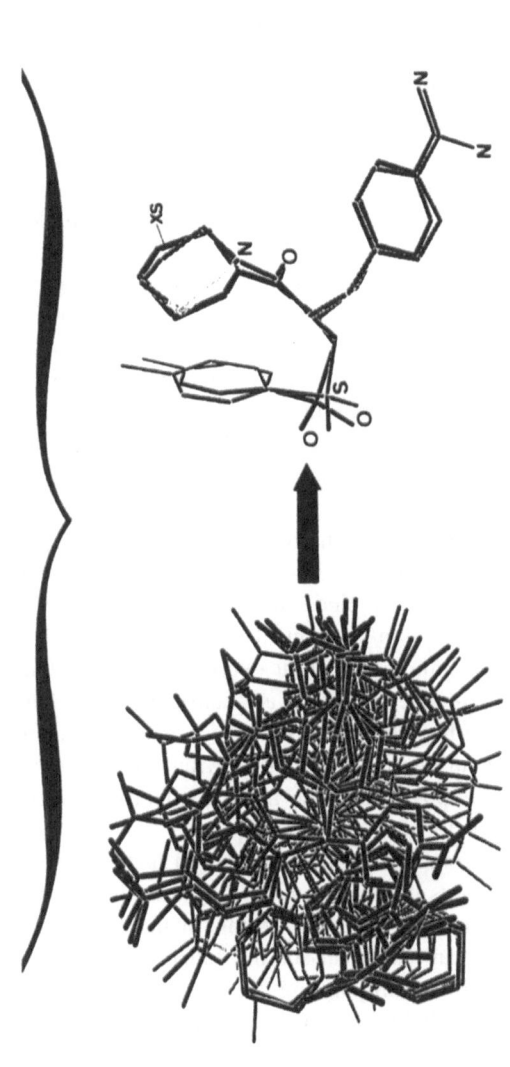

Fig. 6. A conformational analysis, performed with the program MIMUMBA, starts with the 3D structure of a molecule (here: thrombin inhibitor tapap) in an arbitrarily given conformation, e.g., generated by a conversion program such as CORINA or CONCORD. It then analyzes the various torsional fragments and assigns the rotatable bonds to the most similar entries from an internal torsional library. In a combinatorial approach, conformations are generated selecting preferentially those angles that correspond to the experimentally observed conformational preferences. Among a set of 144 generated conformers of tapap one could be found that closely approximates the conformation adopted at the binding site of thrombin (as confirmed crystallographically)

well placed and efficiently distributed in that part of conformation space also accessible to receptor-bound ligands (Klebe 1994a, 1995).

5.8 Computational Methods
Exploiting Conformational Preferences

The information about conformational preferences has been incorporated into an automatic procedure to generate multiple conformations relevant for the structural requirements at the protein binding site (Klebe and Mietzner 1994). In the program MIMUMBA, a ligand is analyzed by assigning rotatable bonds. It is then split into rigid fragments or flexible ring portions. For the latter, low energy conformations are taken from a library of predefined ring templates retrieved from crystal data.

Dihedral angles for rotatable bonds in open-chain portions are selected according to the preferences observed in the corresponding torsional libraries (Fig. 6). In consecutive steps, the program composes the entire molecule from individual torsional fragments, using preferably those open-chain torsion angles that are more frequently found in crystal structures. While merging all fragments in their various conformations, an elaborate ranking system is applied. In addition, a "penalty" is assigned if fragments overlap. In a subsequent geometry optimization, performed in torsion space, unfavorable steric overlap is removed. Only van der Waals and torsion-angle potentials are considered. The latter potentials are derived from the statistically observed torsion angle distributions. They ensure that during minimization, individual torsion angles are maintained as close as possible to the most frequently occurring values in experimental structures.

The approach for the comparison of molecules in terms of shape associated Gaussian functions has been extended by the conformation model in MIMUMBA (Klebe et al. 1994). In this method, molecular similarity is quantified by an analytical scoring function. In order to account for conformational flexibility during the superposition process, this function has been expanded by the empirical torsion-angle potentials used in MIMUMBA. Since this strategy only allows for a local optimization, an initial orientation and a start conformation are required. For this, multiple conformers are generated by a conformational search with MIMUMBA and then subjected to a mutual superposition. This model has also been implemented in the docking tool FlexX (Rarey et al. 1996), providing a fast and efficient mapping of conformation space.

5.9 Hydrogen-Bond Acceptors in Competitive Situations: Which Is the Better Acceptor?

The design of novel protein ligands often involves functional-group replacements. Such bioisosteric replacements are performed to alter a molecular framework while retaining hydrogen-bonding properties or to enhance hydrogen bonding in a well-planned and tailored fashion. In this context, a choice must often be taken as to which is, e.g., the better hydrogen-bond acceptor in two different functional groups or deviating environments? This question is rather complex and more difficult to answer than it might appear at first glance. The knowledge of pK_a values of the compounds involved might help, but these values are often not available or cannot be reliably computed. Partial charges are often used to assess hydrogen-bonding capabilities; however, simple correlations are not sufficient to explain observed differences or trends between different compounds.

Statistical analyses of crystal packings containing molecules with the fragment of interest have shed some light on the differences in relative hydrogen-bonding ability, e.g., of acceptors (Böhm et al. 1996). Examples where oxygen and nitrogen should be able to compete as acceptors in forming hydrogen bonds have been selected for analysis: oxazole, isoxazole, methoxypyridines or oxime ethers. In these competitive situations no or only very few hydrogen bonds are formed with oxygen, whereas nitrogen is frequently involved. In these examples, oxygen is bound to two non-hydrogen atoms, at least one of which is formally assigned an sp^2-type hybridization.

In order to further validate the observation that oxygen is a very weak H-bond acceptor in this environment, crystal packing data of furanes, enol ethers, ethers, and esters have been studied. Furans and enol ethers emerge to be rare hydrogen-bond acceptors. For esters, two different oxygen atoms are in competitive situations. The ether-type oxygen is rarely involved in hydrogen bonds, while the carbonyl oxygen frequently accepts such interactions. In aliphatic ethers, data indicate oxygen to be a fairly good hydrogen-bond acceptor. These data can be

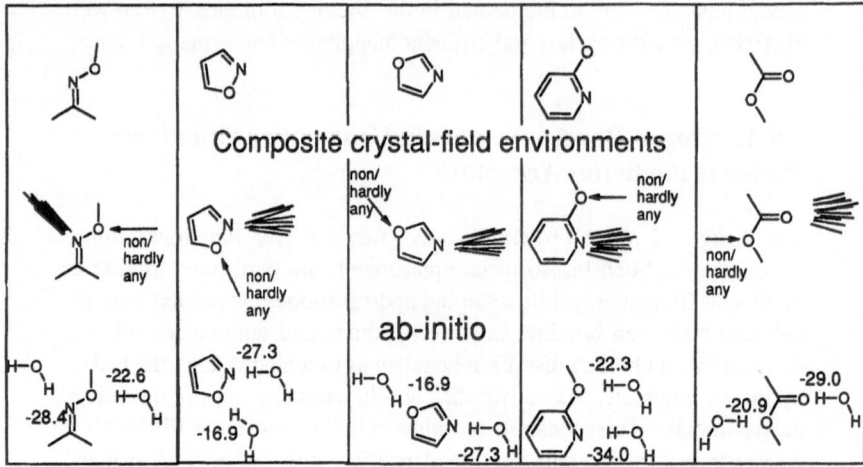

Fig. 7. For different functional groups presenting oxygen and nitrogen atoms in potentially competitive hydrogen-bonding situations, the oxygen, bound to two non-hydrogen atoms of which at least one is formally assigned to an sp^2-type hybridization, shows hardly any tendency to be involved in H-bonding. This behavior is confirmed either by analyzing H-bonding properties in crystal packing or by performing high-level ab initio calculations between the functional group under consideration and a water probe (all energies in kJ/mol)

compared with high level ab initio calculations. The heat of formation of complexes of representative molecules containing the functional groups of interest and a water molecule have been computed. It is self-evident that if a hydrogen bond between a particular functional group and a water molecule is computed to be weak, the statistical evaluation of crystal data shows low involvement frequencies or reduced probabilities for H-bond formation (Fig. 7).

The evaluation technique described can be applied generally and permits estimation of the H-bonding capabilities of molecules (in the new IsoStar tool, available from CCDC, statistics on crystal data are given together with results from ab initio calculations). In drug design, functional group replacements are undertaken for various reasons, e.g., to improve the binding affinity of a ligand. The present correlations provide some guidelines for estimating how successful such replacements can be expected to be. With respect to the prediction of ligand binding either in docking or de novo design, one is often faced with the problem that there may be several ways for a ligand to satisfy donor-functional groups of the protein at the binding site. The described correlations can help to discriminate among prospective groups (e.g., oxygen atoms adjacent to sp^2 centers are unlikely to be involved in structure determining hydrogen bonds) and thus control molecular recognition and binding properties. This will especially be the case if more advantageous acceptors are present in a ligand or could be conceived in a de novo design approach.

5.10 Conclusion

Complex formation between a protein and a ligand is governed by non-covalent bonding interactions that follow the principles of molecular recognition. Once understood, these principles can be translated into rules and then exploited for the design of novel ligands. Over the last decade, we have witnessed an exponential increase of knowledge about the intra- and intermolecular 3D geometry of organic molecules and proteins, information that has emerged mainly from crystal data. In a crystal possessing long-range periodicity, the mutual short-range recognition of molecules produces an ordered arrangement. Accordingly, important information about general recognition patterns ("supra-molecular synthons") can be derived from these structures. Such knowledge from crystal data has been implemented in computational tools used in drug design, especially for the prediction of ligand-protein interactions and the comparison of protein-bound ligands. They provide solutions for docking, de novo design, conformational analysis, molecular comparison, structural superposition and isosteric functional group replacement. These are only some aspects of interest in the field of drug design. Many more structural problems await to be accessed via knowledge-based approaches.

References

Allen FH, Kennard O, Taylor R (1983) Systematic analysis of structural data as a research technique in organic chemistry. Acc Chem Res 16:146–153

Bernstein FC, Koetzle TF, Williams GJB, Meyer EF Jr, Brice MD, Rodgers JR, Kennard O, Shimanouchi T, Tasumi M (1977) The protein data bank: a computer-based archival file for macromolecular structures. J Mol Biol 112: 535–542

Böhm H-J (1992a) The computer program LUDI: a new method for the de novo design of enzyme inhibitors. J Comput Aided Mol Design 6:61–78

Böhm H-J (1992b) LUDI: rule-based automatic design of new substituents for enzyme inhibitor leads. J Comput Aided Mol Design 6:593–606

Böhm H-J (1994) The development of a simple empirical scoring function to estimate the binding constant for a protein-ligand complex of known three-dimensional structure. J Comput Aided Mol Design 8:243–256

Böhm H-J (1998) Prediction of binding constants of protein ligands: a fast method for the prioritization of hits obtained from de novo design or 3D database search programs. J Comput Aided Mol Design 12:309–323

Böhm H-J, Klebe G (1996) What can we learn from molecular recognition in protein-ligand complexes for the design of new drugs? Angew Chem Int Ed Engl 35:2588–2614

Böhm H-J, Brode S, Hesse U, Klebe G (1996) Oxygen and nitrogen in competitive situations: which is the hydrogen-bond acceptor? Chem Eur J 2:1509–1513

Bruno IJ, Cole JC, Lommerse JPM, Rowland RS, Taylor R, Verdonk ML (1997) IsoStar: a library of information about nonbonded interactions. J Comput Aided Mol Design 11:525–537

Desiraju GR (1995) Supramolecular synthons in crystal engineering – a new organic synthesis. Angew Chem Int Ed Engl 34:2311–2327

Dunitz JD (1991) Phase transitions in molecular crystals from a chemical view point. Pure Appl Chem 63:177–185

Hemm K, Hendlich M, Aberer K (1995) Constituting a receptor ligand information base from quality-enriched data. In: Proceedings from the third international conference on intelligent systems for molecular biology. Rawling CJ, AAAI Press, Menlo Park, Calif, pp 170–178

Hol W, Wierenga RK (1984) The x-helix dipole and the binding of phosphate groups of coenzymes by proteins. In: Horn AS, DeRanter CJ (eds) X-ray crystallography and drug action. Clarendon, Oxford

Kearsley SK, Smith GM (1990) An alternative method for the alignment of molecular structures: maximizing electrostatic and steric overlap. Tetrahed Comput Methods 3:615–633

Klebe G (1994a) Structure correlation and ligand/receptor interactions. In: Bürgi H-B, Dunitz JD (eds) Structure correlation. VCH, Weinheim, pp 543–603

Klebe G (1994b) The use of composite crystal-field environments in molecular recognition and the 'de-novo' design of protein ligands. J Mol Biol 237:212–235

Klebe G (1995) Toward a more efficient handling of conformational flexibility in computer-assisted modelling of drug molecules. Perspect Drug Discov Design 3:85–105

Klebe G, Diederich F (1993) A comparison of the crystal packing in benzene with the geometry seen in crystalline cyclophane-benzene complexes – guidelines for rational receptor design. Philos Trans R Soc Lond A 345:37–48

Klebe G, Mietzner T (1994) A fast and efficient method to generate biologically relevant conformations. J Comput Aided Design 8:583–606

Klebe G, Mietzner T, Weber F (1994) Different approaches toward an automatic structural alignment of drug molecules: applications to sterol mimics, thrombin and thermolysin inhibitors. J Comput Aided Design 8:751–778

Klebe G, Mietzner T, Weber F (1998) Methodological developments and strategies for a fast flexible superposition of drug-size molecules. J Comput Aided Mol Design (in press)

Lehn J-M (1973) Design of organic complexing agents, strategies towards properties. Struct Bonding (Berlin) 16:1–69

Lehn J-M (1988) Supramolecular chemistry – scope and perspectives. Molecules, supermolecules, and molecular devices (Nobel lecture). Angew Chem Int Ed Engl 27:89–112

Nicklaus MC, Wang S, Driscoll JS, Milne GWA (1995) Conformational changes of small molecules binding to proteins. Bioorgan Med Chem 3:411–428

Nissink JWM, Verdonk M L, Kroon J, Mietzner T Klebe G (1997) Superposition of molecules: electron density fitting by application of Fourier transforms. J Comput Chem 18:638–645

Rarey M, Kramer B, Lengauer T, Klebe G (1996) A fast flexible docking method using an incremental construction algorithm. J Mol Biol 261:470–489

Ricketts EM, Bradshaw J, Hann M, Hayes F, Tanna N, Ricketts DM (1993) Comparison of conformations of small molecule structures from the protein data bank with those generated by concord, cobra, ChemDBS-3D, and converter and those extracted from the Cambridge structural database. J Chem Infect Comput Sci 33:905–925

Sussman J L, Harel M, Frolow F, Oefner C, Goldman A, Toker L, Silman I
(1991) Atomic structure of acetylcholinesterase from Torpedo California: a
prototypic acetylcholine-binding protein. Science 253:872–879

6 Sequence-Specific Recognition of DNA and Control of Gene Expression by Oligonucleotide-Intercalator Conjugates

C. Hélène, C. Giovannangeli, J.-S. Sun, and T. Garestier

6.1   Introduction 119
6.2   Sequence-Specific Triple-Helix Formation
        on Double-Helical DNA 120
6.3   Transcription Inhibition
        by Triple Helix-Forming Oligonucleotides 123
6.4   Accessibility of DNA in Cell Nuclei
        to Triple Helix-Forming Oligonucleotides 125
6.5   Clamp Oligonucleotides 127
6.6   Conclusion 128
References 130

6.1 Introduction

Gene expression can be controlled with oligonucleotides according to several strategies (for review see Hélène 1994). *Antisense* oligonu-cleotides bind to complementary sequences on messenger RNAs and inhibit translation of the message into the coded protein. *Ribozymes* are also targeted to messenger RNAs (or viral RNAs) and induce a catalytic cleavage of the recognized RNA, thereby inhibiting translation of the mRNA (or expression of the viral RNA). An oligonucleotide *decoy* can be used to sequester a transcription factor and to control the expression of genes that are regulated by this transcription factor. Several genes are expected to respond to the oligonucleotide decoy due to the involvement of each transcription factor in the regulation of multiple genes. Oligonucleotide *aptamers* can be selected to target proteins involved at any step of gene control and expression.

Control of gene transcription can be achieved with *antigene* oligonucleotides that bind to double-helical DNA to form a local triple helix (Le Doan et al. 1987; Moser and Dervan 1987). In contrast, PNAs (peptide nucleic acids) may inhibit transcription by strand invasion of a double-helical template (Nielsen et al. 1994). The targeted sequence may be located in the promoter or enhancer region of the gene or within the transcribed portion. A triple helix can also be formed on a single-stranded nucleic acid by *clamp* (Giovannangeli et al. 1991, 1993) or *circular* oligonucleotides (Kool 1991; Wang and Kool 1995). When the target is an RNA, these oligonucleotides are expected to inhibit translation (messenger RNA) or reverse transcription (viral RNA). This review will deal with triple helix-forming oligonucleotides and their gene regulatory activities.

6.2 Sequence-Specific Triple-Helix Formation on Double-Helical DNA

Triple helix formation involves the recognition of Watson-Crick base pairs by hydrogen bonding interactions within the major groove of the double helix (Thuong and Hélène 1993). Oligonucleotides and oligonucleotide analogues can wind around the double helix; their orientation is dependent on base sequence (Fig. 1). Only purine bases in the target DNA can form two hydrogen bonds with the oligonucleotide; a regular winding of the third strand oligonucleotide requires that all purines are on the same strand of DNA.

Fig. 1. Base triplets formed by natural nucleosides with Watson-Crick T.A and C.G base pairs. *The column on the left-hand side* corresponds to the *Hoogsteen* configuration. The third strand runs parallel to the oligopurine target. *The column on the right-hand side* corresponds to the *reverse Hoogsteen* configuration. The third strand is antiparallel to the oligopurine strand of the double-helical target (Sun et al. 1996)

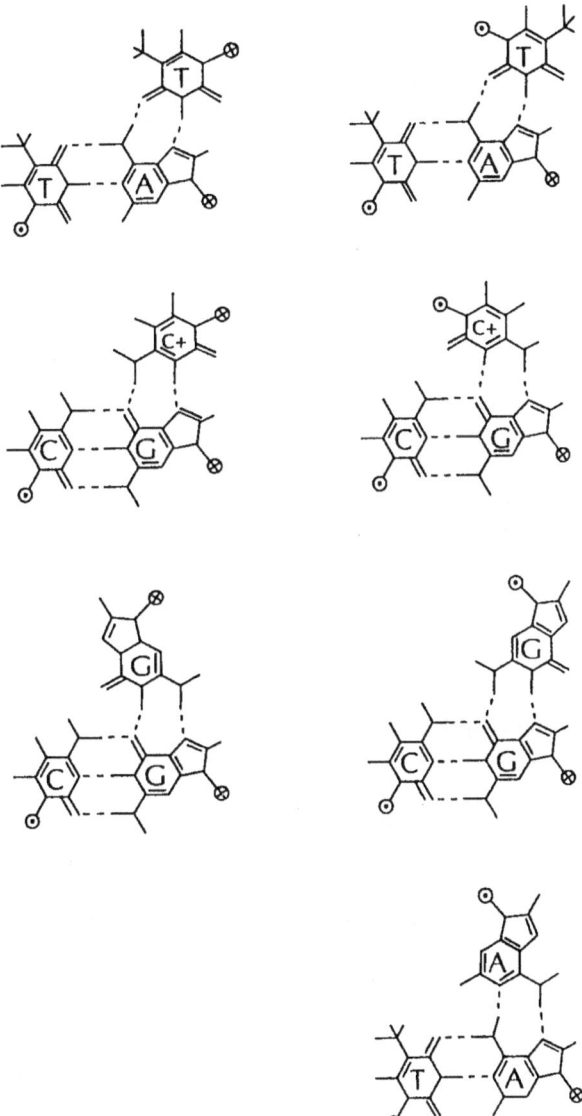

Fig. 1. Legend see p. 120

Recognition of the purines in T.A and C.G base pairs may be achieved by T and protonated C (C⁺), respectively, by forming Hoog-steen hydrogen bonds (as originally described by Hoogsteen in 1963). Pyrimidic oligonucleotides adopt a parallel orientation with respect to the oligopurine sequence. The two base triplets T.AxT and C.GxC⁺ are isomorphous, i.e., the oligopyrimidine winds without any distortion of its backbone around the targeted double-helical sequence. The require-ment for cytosine protonation to form a stable C.GxC⁺ base triplet makes the stability of triple helices pH-dependent for (C,T)-containing oligonucleotides. However, triple helices can be observed at pH 7 if most cytosines have thymines and not cytosines as neighbors.

Alternatively, the purines of T.A and C.G base pairs can be recog-nized by A and G, respectively. A purinic oligonucleotide binds in an antiparallel orientation with respect to the target oligopurine sequence. The two base triplets T.AxA and C.GxG are not isomorphous; therefore an adjustment of the backbone conformation is required to form a triple helix.

The parallel orientation of (T,C)-containing oligonucleotides and the antiparallel one of (A,G)-containing oligonucleotides assumes that all nucleotides adopt an *anti* conformation (Beal and Dervan 1991). Such orientations have been observed in all experiments reported to date. A *syn* conformation of the nucleosides would lead to a reverse orientation. It should be noted that T and C⁺ can form base triplets with T.A and C.G base pairs, respectively, in a *reverse Hoogsteen* configuration that should lead to an antiparallel orientation of the (T,C)-containing third strand. This has never been observed with natural oligonucleotides, probably because T.AxT and C.GxC⁺ base triplets are not isomorphous and the free energy of base triplet formation and stacking may be less favorable in the reverse Hoogsteen as compared to the Hoogsteen con-figuration.

Oligonucleotides synthesized with G's and T's can also form triple helices with an oligopyrimidine•oligopurine sequence of double-helical DNA. The orientation of the (G,T)-containing oligonucleotide depends on base composition (number of 5' GpT 3' and 5' TpG 3' steps, length of G and T tracts). Parallel and antiparallel orientations involve Hoog-steen and reverse Hoogsteen configurations of the CGxG and TAxT base triplets, respectively (Sun et al. 1996).

In order for the third strand oligonucleotide to wind smoothly around the major groove of DNA, all purines of the target sequence must be on the same strand of the double helix. Otherwise the backbone of the third strand would have to cross the major groove at the site where a pyrimidine interrupts the oligopurine tract. However, it is possible to recognize a pyrimidine within an oligopurine sequence by a base forming a single hydrogen bond with the pyrimidine base. This possibility has been exemplified by introducing a guanine in a (C,T)-oligonucleotide to recognize a thymine in an oligopurine sequence. The energy loss associated with this non-canonical ATxG base triplet depends on the flanking base triplets (Kiessling et al. 1992). It is also possible to enhance the binding energy by attaching an intercalating agent at the site facing the interruption of the oligopurine sequence (Zhou et al. 1995).

The recognition of two oligopurine sequences alternating on the two strands of the DNA double helix can be achieved by two oligonucleotides linked to each other by a linker whose length and nature depends on the bases in the third strand and the site (5' PuPy 3' or 5' PyPu 3') where the third strand crosses the major groove (see Sun 1995, for review).

6.3 Transcription Inhibition by Triple Helix-Forming Oligonucleotides

The specificity of recognition of a double-helical target by an oligonucleotide provides the basis of the so-called "antigene" strategy to inhibit gene expression at the transcriptional level (Hélène 1991, 1994). When the target is located within the control region (promoter, enhancer), the bound oligonucleotide may inhibit transcription factor binding. When the oligonucleotide binds downstream of the transcription start site it may inhibit the elongation step of the transcription process.

The possibility of inhibiting transcription by a (G,T)-containing oligonucleotide was first described in an in vitro transcription system (Cooney et al. 1988). Binding of an oligonucleotide to a transcription factor binding site competes with the binding of the regulatory protein and, thereby, modulates transcription initiation (Cooney et al. 1988; Maher et al. 1992; Grigoriev et al. 1992; Ing et al. 1993). Several in vitro transcription systems have been used to demonstrate this competitive

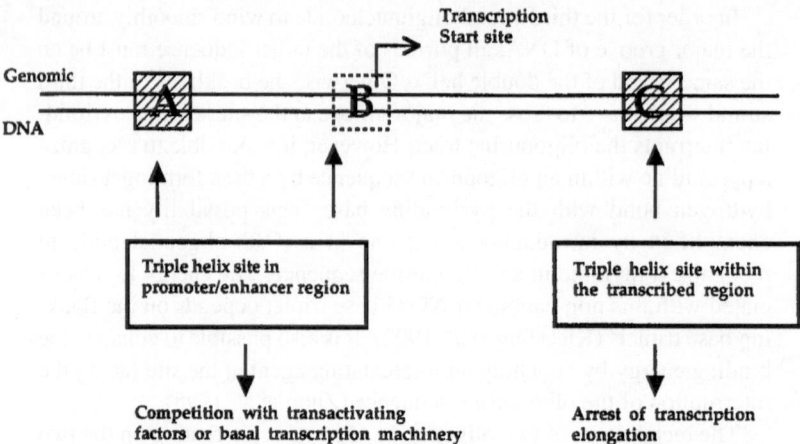

Fig. 2. Location of potential binding sites of triplex-forming oligonucleotides on genomic DNA. In *A* the oligonucleotide competes with the binding of transcriptional activators, in *B* with the basal transcription machinery; in *C* the oligonucleotide may arrest the transcription machinery during the elongation step

inhibition (Fig. 2). However, the elongation process is much more difficult to inhibit because the stability of the triple-helical complex is usually not sufficient to arrest the transcription machinery once it is launched on its double-helical template. Two strategies have been developed to achieve transcription arrest:

1. The oligonucleotide can be covalently attached to an intercalating agent (Fig. 3). The oligonucleotide-intercalator conjugate binds more tightly to its target DNA due to the additional binding energy provided by intercalation at the triplex-duplex junction or within the triple-helical region (Sun et al. 1989; Giovannangeli et al. 1996; Silver et al. 1997).

2. Chemical modifications of the oligonucleotide may provide the analogue with a tighter binding affinity. PNAs do bind tightly to double-helical DNA but they involve a strand-displacement reaction where two PNA molecules bind to an oligopurine sequence on *one* strand of the double helix, forming a local triple helix (Nielsen et al. 1994). Several other chemical modifications have been tested for

their ability to form triple helices (Escudé et al. 1993). 2'-*O*-Methyl pyrimidine oligonucleotides form more stable complexes than DNA and RNA oligonucleotides. A (C,T)-containing RNA binds more tightly than the corresponding DNA oligonucleotide to a DNA double helix. In contrast, neither 2'-*O*-methyl nor RNA *purine* oligonucleotides form stable triple helices on a DNA double helix in contrast to a DNA oligonucleotide. Of all the chemical modifications tested so far, N3'→P5' phosphoramidate linkages confer upon *pyrimidine* oligonucleotides the strongest binding described to date (Escudé et al. 1996). *Purine* oligophosphoramidates do not appear to form stable triple helices.

6.4 Accessibility of DNA in Cell Nuclei to Triple Helix-Forming Oligonucleotides

Oligonucleotide-intercalator conjugates, PNA and oligophosphoramidates have been shown to inhibit transcription in vitro in a sequence-specific manner. The data available within cells are more scarce. Very few studies provide evidence that the effect observed on gene expression is due to oligonucleotide binding to DNA to form a triple-helical complex. In some cases the observed effect on transcription might be due to binding of the oligonucleotide to a transcription factor rather than to DNA. When the gene of interest is carried by a plasmidic vector it is possible to provide evidence for the involvement of triple helix formation in the inhibition of gene transcription by introducing mutations in the target sequence. In most cases, however, especially for an endogenous gene, it is difficult to construct a mutant of the target sequence. The control experiments rely upon modifications of the oligonucleotide sequence even though there are problems associated with this type of control since all potential interactions of the oligonucleotide (including self association) may change upon sequence modification.

One of the main questions raised by the development of the antigene strategy in vivo is the accessibility of the target sequence within the chromatin structure in the cell nucleus. To provide evidence for accessibility, we have developed a strategy based upon using oligonucleotide-psoralen conjugates (Fig. 3). When such a conjugate forms a triple-helical complex with DNA, the psoralen moiety can be cross-linked to one

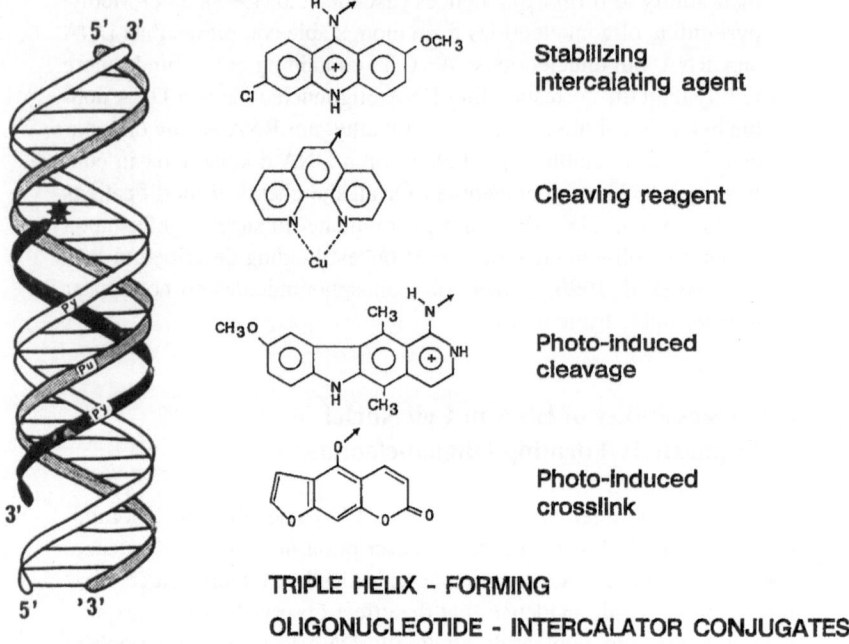

Fig. 3. *Left* Schematic representation of a triple-helical complex where an oligonucleotide (*black ribbon*) wraps around the major groove of the double helix. The oligonucleotide can be covalently attached (*star*) to an intercalating agent that (a) stabilizes the triplex (Sun et al. 1989), (b) induces chemical cleavage of the target site (François et al. 1989), (c) photo-induces cleavage of the target double helix (Perrouault et al. 1990), and (d) can be used to crosslink the two strands of DNA under UV irradiation (Takasugi et al. 1991; Giovannangeli et al. 1997)

or both strands of the double helix upon UV irradiation (Takasugi et al. 1991). If the site of triple helix formation and cross-linking overlaps a restriction site, it is possible to reveal the inhibition of restriction enzyme cleavage at this particular site by using probes that overlap the targeted DNA region. The absence of inhibition at other restriction sites for the same enzyme provides an internal control of the sequence specificity of the cross-linking reaction and, therefore, of triple helix formation. This strategy has been used to demonstrate the accessibility of the proviral HIV sequence in chronically infected cells (Giovannangeli et al. 1997). This might not be true of all targeted sequences due to the nucleosomal structure of chromatin. If the oligonucleotide interacts with a sequence where transcription factors bind to activate transcription, it is likely that the oligonucleotide may also have access to its target sequence in a similar fashion to transcription factors. Kinetic parameters might play an important role inasmuch as some triple-helical complexes exhibit a slow rate of formation as compared to protein binding (Maher et al. 1990; Rougée et al. 1992).

Oligonucleotide-psoralen conjugates have been used to induce site-specific mutations on plasmids. These mutations are located at the specific site where psoralen cross-linking is induced by UV irradiation after triple helix formation. They clearly indicate that the target site has been reached by the oligonucleotide within cells. However, until now the target sites have been limited to plasmidic vectors and not to endogenous genes (Wang et al. 1995; Sandor and Bredberg 1994; Raha et al. 1996).

6.5 Clamp Oligonucleotides

An oligopurine sequence on a single-stranded nucleic acid can be recognized by a complementary (antisense) oligonucleotide. The short double helix with an oligopyrimidine•oligopurine sequence can, in turn, be recognized by a third strand oligonucleotide to form a triple helix. The two oligonucleotides can be linked to each other to form a unique molecule that can clamp the target sequence on the single-stranded template (Giovannangeli et al. 1991; Fig. 4). The nature of the third strand (oligopyrimidine, oligopurine or (G,T)-oligonucleotide) determines its orientation with respect to the oligopurine target sequence. Therefore, the linker between the antisense and the "antigene" portions will join a 3'- to a 3'-end or a 5'- to a 5'-end (for an antiparallel orientation of the third strand) or a 3'- to a 5'-end (for a parallel orientation). In the last case, a circular oligonucleotide can be synthesized (Kool 1991). For (mostly) entropic reasons the circular oligonucleotide will bind more tightly than the clamp oligonucleotide, which in turn binds much more tightly than two separate oligonucleotides (at least in the micromolar range of concentrations).

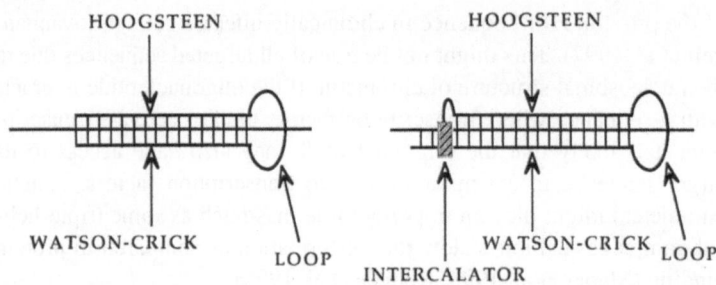

Fig. 4. Oligonucleotide clamps formed of two portions connected by a loop; they form both Watson-Crick and Hoogsteen hydrogen bonds with the single-stranded target. Attachment of an intercalator to one end of the clamp oligonucleotide stabilizes the complex (provided the Watson-Crick portion is made longer than the Hoogsteen portion; Giovannangeli et al. 1991, 1993)

Clamp oligonucleotides have been shown to inhibit primer extension by DNA polymerase on a single-stranded template under conditions where antisense oligonucleotides are devoid of any inhibitory activity (Giovannangeli et al. 1993). They can also arrest reverse transcription on a single-stranded RNA template. We have recently shown (C. Giovannangeli et al., unpublished results) that a clamp oligonucleotide is able to block HIV infection of CD4-positive cells at an early step after infection, most likely reverse transcription. No proviral DNA is detected after viral infection. An antisense oligonucleotide targeted to the same sequence exhibited no inhibition.

Clamp oligonucleotides can be covalently attached to an intercalating agent (Fig. 4). If the antisense portion is made a little longer than the third strand portion, intercalation can lock the complex in place on the single-stranded target (Giovannangeli et al. 1993).

6.6 Conclusion

Triple helix formation represents an alternative to antisense oligonucleotides to control gene function. Antigene oligonucleotides targeted to the DNA double helix can inhibit transcription. Clamp oligonucleotides targeted to a viral sequence can inhibit reverse transcription. They might also inhibit translation of a messenger RNA (even though there are no data yet available on translation inhibition). Circular oligonucleotides forming a triple helix with a single-stranded template might also be useful in both approaches. The strand displacement reaction observed with PNAs (which involves triple helix formation by two PNAs on one of the two DNA strands) might represent an alternative to antigene oligonucleotides which bind to DNA without any opening of the double helix. Further experiments with oligonucleotide analogues that form stable triple helices (e.g., oligophosphoramidates) will tell us whether the antigene or clamp strategies can be applied to biologically relevant in vivo situations.

Antisense oligonucleotides have reached the stage of clinical trials in several pathological disorders. The information gained on bioavailability, pharmacokinetics, delivery, routes of administration, etc., will be useful in any development of antigene oligonucleotides. Whether there is any advantage targeting the gene rather than its messenger RNA (or pre-mRNA) remains to be determined in each particular case. Nuclease-resistant analogues (such as $N3' \rightarrow P5'$ oligophosphoramidates) could have long-lasting effects on gene transcription. The lower number of targets (two alleles for each gene) as compared to messenger RNAs might be an obvious advantage, especially for oligonucleotide analogues that do not bind strongly to cellular proteins. This should allow us to obtain a biological response at rather low oligonucleotide concentrations provided the target sequence is accessible within the chromatin structure of cell nuclei as recently demonstrated (Guieysse-Peugeot et al. 1996; Giovannangeli et al. 1997). In addition, there might be proteins within cells that bind strongly to triple-helical complexes formed upon binding of antigene oligonucleotides to their target DNA sequence. Proteins with these characteristics have been recently described (Kiyama et al. 1991; Guieysse-Peugeot et al. 1997).

Target sequences for antigene oligonucleotides remain limited to oligopyrimidineoligopurine tracts of double-helical DNA. The design of nucleoside analogues or modifications of oligonucleotides involving, e.g., the insertion of intercalating agents should allow us to extend the range of triple helix-forming DNA sequences. Together with minor groove-binding ligands with an increased range of sequence-specific recognition (Gottesfeld et al. 1997; White et al. 1998), major groove-specific ligands such as antigene oligonucleotides provide a new way of controlling gene transcription in vivo. The parallel development of a gene therapy approach based on triple helix formation (Rininsland et al. 1997; Shevelev et al. 1997) opens new avenues for controlling gene expression in pathological disorders.

References

Beal PA, Dervan PB (1991) Second structural motif for recognition of DNA by oligonucleotide-directed triple-helix formation. Science 251:1360–1363

Cooney M, Czernuszewicz G, Postel EH, Flint SJ, Hogan ME (1988) Site-specific oligonucleotide binding represses transcription of the human c-myc gene in vitro. Science 241:456–459

Escudé C, François J-C, Sun J-S, Ott G, Sprinzl M, Garestier T, Hélène C (1993) Stability of triple helices containing RNA and DNA strands: experimental and molecular modeling studies. Nucleic Acids Res 21:5547–5553

Escudé C, Giovannangeli C, Sun JS, Lloyd DH, Chen JK, Gryaznov S, Garestier T, Hélène C (1996) Stable triple-helices formed by oligonucleotide N3'→P5' phosphoramidates inhibit transcription elongation. Proc Natl Acad Sci USA 93:4365–4369

François J-C, Saison-Behmoaras T, Barbier C, Chassignol M, Thuong NT, Hélène C (1989) Sequence-specific recognition and cleavage of duplex DNA via triple-helix formation by oligonucleotides covalently linked to a phenanthroline-copper chelate. Proc Natl Acad Sci USA 86:9702–9706

Giovannangeli C, Montenay-Garestier T, Rougée M, Chassignol M, Thuong NT, Hélène C (1991) Single-stranded DNA as a target for triple-helix formation. J Am Chem Soc 113:7775–7777

Giovannangeli C, Thuong NT, Hélène C (1993) Oligonucleotide clamps arrest DNA synthesis on a single-stranded DNA target. Proc Natl Acad Sci USA 90:10013–10017

Giovannangeli C, Perrouault L, Escudé C, Thuong N, Hélène C (1996) Specific inhibition of in vitro transcription elongation by triplex-forming oligonucleotide-intercalator conjugates targeted to HIV proviral DNA. Biochemistry 35:10539–10548

Giovannangeli C, Diviacco S, Labrousse V, Gryaznov S, Charneau P, Hélène C (1997) Accessibility of nuclear DNA to triplex-forming oligonucleotides: the integrated HIV-1 provirus as a target. Proc Natl Acad Sci USA 94:79–84

Gottesfeld JM, Neely L, Trauger JW, Baird EE, Dervan PB (1997) Regulation of gene expression by small molecules. Nature 387:202–205

Grigoriev M, Praseuth D, Robin P, Hemar A, Saison-Behmoaras T, Dautry-Varsat A, Thuong NT, Hélène C, Harel-Bellan A (1992) A triple helix-forming oligonucleotide-intercalator conjugate acts as a transcriptional repressor via inhibition of NF KB binding to Interleukin-2 receptor α-subunit regulatory sequence. J Biol Chem 267:3389–3395

Grigoriev M, Praseuth D, Guieysse AL, Robin P, Thuong NT, Hélène C, Harel-Bellan A (1993) Inhibition of interleukin-2 receptor α-subunit gene expression by oligonucleotide-directed triple helix formation. C R Acad Sci III Sci Vie 316:492–495

Guieysse A-L, Praseuth D, Hélène C (1997) Identification of a triplex DNA-binding protein from human cells. J Mol Biol 267:289–298

Guieysse A-L, Praseuth D, Grigoriev M, Harel-Bellan A, Hélène C (1996) Detection of covalent triplex within human cells. Nucleic Acids Res 24:4210–4216

Hélène C (1991) The anti-gene strategy: control of gene expression by triplex-forming-oligonucleotides. Anti Cancer Drug Design 6:569–584

Hélène C (1994) Control of oncogene expression by antisense nucleic acids. Eur J Cancer 30A:1721–1726

Hoogsteen K (1963) The crystal and molecular structure of a hydrogen-bonded complex between 1-methylthymine and 9-methyladenine. Acta Crystallogr 16:907–916

Ing NH, Beekman JM, Kessler DJ, Murphy M, Jayaraman K, Zendegui JG, Hogan ME, O'Malley BW, Tsai M-J (1993) In vivo transcription of a progesterone-responsive gene is specifically inhibited by a triplex-forming oligonucleotide. Nucleic Acids Res 21:2789–2796

Kiessling LL, Griffin LC, Dervan PB (1992) Flanking sequence effects within the pyrimidine triple-helix motif characterized by affinity cleaving. Biochemistry 31:2829–2834

Kiyama R, Camerini-Otero RD (1991) A triplex DNA-binding protein from human cells: purification and characterization. Proc Natl Acad Sci USA 88:10450–10454

Kool ET (1991) Molecular recognition by circular oligonucleotides: increasing the selectivity of DNA binding. J Am Chem Soc 113:6265–6266

Le Doan T, Perrouault L, Praseuth D, Habhoub N, Decout JL, Thuong NT, Lhomme J, Hélène C (1987) Sequence-specific recognition, photocrosslinking and cleavage of the DNA double helix by an oligo-[α]-thymidylate covalently linked to an azidoproflavine derivative. Nucleic Acids Res 15:7749–7760

Maher LJ III, Dervan PB, Wold BJ (1990) Kinetic analysis of oligodeoxyribonucleotide-directed triple-helix formation on DNA. Biochemistry 29:8820–8826

Maher LJ III, Dervan PB, Wold B (1992) Analysis of promoter-specific repression by triple-helical DNA complexes in a eukaryotic cell-free transcription system. Biochemistry 31:70–81

Moser HE, Dervan PB (1987) Sequence-specific cleavage of double helical DNA by triple helix formation. Science 238:645–650

Nielsen PE, Egholm M, Buchardt O (1994) Peptide nucleic acid (PNA). A DNA mimic with a peptide backbone. Bioconjug Chem 5:3–7

Perrouault L, Asseline U, Rivalle C, Thuong NT, Bisagni E, Giovannangeli C, Le Doan T, Hélène C (1990) Sequence-specific artificial photo-induced endonucleases based on triple helix-forming oligonucleotides. Nature 344:358–360

Raha M, Wang G, Seidman MM, Glazer PM (1996) Mutagenesis by third-strand-directed psoralen adducts in repair-deficient human cells: high frequency and altered spectrum in a xeroderma pigmentosum variant. Proc Natl Acad Sci USA 93:2941–2946

Rininsland F, Johnson TR, Chernicky CL, Schulze E, Burfeind P, Ilan J, Ilan J (1997) Suppression of insulin-like growth factor type I receptor by a triple-helix strategy inhibits IGF-I transcription and tumorigenic potential of rat C6 glioblastoma cells. Proc Natl Acad Sci USA 94:5854–5859

Rougée M, Faucon B, Mergny JL, Barcelo F, Giovannangeli C, Garestier T, Hélène C (1992) Kinetics and thermodynamics of triple helix formation: effects of ionic strength and mismatches. Biochemistry 31:9269–9278

Sandor Z, Bredberg A (1994) Repair of triple helix directed psoralen adducts in human cells. Nucleic Acids Res 22:2051–2056

Shevelev A, Burfeind P, Schulze E, Rininsland F, Johnson TR, Trojan J, Chernicky CL, Hélène C, Ilan J, Ilan J (1997) Potential triple helix-mediated inhibition of IGF-I gene expression significantly reduces tumorigenicity of glioblastoma in an animal model. Cancer Gene Ther 4:105–112

Silver GC, Sun J-S, Nguyen CH, Boutorine AS, Bisagni E, Hélène C (1997) Stable triple-helical DNA complexes formed by benzopyridoindole- and benzopyridoquinoxaline-oligonucleotide conjugates. J Am Chem Soc 119:263–268

Sun J-S (1995) Rational design of switched triple helix-forming oligonucleotides: extension of sequences for triple helix formation. In: Pullman A et al (eds) Modelling of bimolecular structures and mechanisms. Kluwer Academic, Dordrecht, pp 267–288

Sun J-S, François JC, Montenay-Garestier T, Saison-Behmoaras T, Roig V, Thuong NT, Hélène C (1989) Sequence-specific intercalating agents: intercalation at specific sequences on duplex DNA via major groove recognition by oligonucleotide-intercalator conjugates. Proc Natl Acad Sci USA 86:9198–9202

Sun J-S, Garestier T, Hélène C (1996) Oligonucleotide directed triple helix formation. Curr Opin Struct Biol 6:327–333

Takasugi M, Guendouz A, Chassignol M, Decout JL, Lhomme J, Thuong NT, Hélène C (1991) Sequence-specific photo-induced cross-linking of the two strands of double-helical DNA by a psoralen covalently linked to a triple helix-forming oligonucleotide. Proc Natl Acad Sci USA 88:5602–5606

Thuong NT, Hélène C (1993) Sequence-specific recognition and modification of double-helical DNA by oligonucleotides. Angew Chem Int Ed Engl 32:666–690

Wang S, Kool T (1995) Relative stabilities of triple helices composed of combinations of DNA, RNA and 2'-O-methyl-RNA backbones: chimeric circular oligonucleotides as probes. Nucleic Acids Res 23:1157–1164

Wang G, Levy DD, Seidman MN, Glazer PM (1995) Targeted mutagenesis in mammalian cells mediated by intracellular triple helix formation. Mol Cell Biol 15:1759–1768

White S, Szewczyk JW, Turner JM, Baird EE, Dervan PB (1998) Recognition of the four Watson-Crick base pairs in the DNA minor groove by synthetic ligands. Nature 391:468–471

Zhou B-W, Puga E, Sun J-S, Garestier T, Hélène C (1995) Stable triple helices formed by acridine-containing oligonucleotides with oligopurine tracts of DNA interrupted by one or two pyrimidines. J Am Chem Soc 117:10425–10428

7 Combinatorial Nucleic Acid Libraries: The New World of Aptamers and Ribozymes

M. Famulok

7.1   Aptamers for Small Molecules: Structure and Function 135
7.2   Functional Aptamers for Proteins and Their Application
        in Biotechnology, Molecular Medicine, and Diagnostics 138
7.2.1   Prion Specific RNA Aptamers 138
7.2.2   Functional Nuclease Resistant Aptamers 140
7.2.3   Functional Aptamers In Vivo 142
7.3   Catalytic Nucleic Acids Selected from Combinatorial Libraries .. 143
References 145

7.1 Aptamers for Small Molecules: Structure and Function

In vitro selection of nucleic acids is a combinatorial technique. Nucleic acids are particularly suited for combinatorial selection approaches because they can fold into well defined secondary, tertiary and quarternary structures and because they can easily be amplified by the polymerase chain reaction (PCR) or in vitro transcription. Thus, phenotype (the three-dimensional fold) and genotype are directly linked in the same molecule. A considerable number of aptamer structures have been solved recently by multidimensional nuclear magnetic resonance (NMR). These studies, as well as the first crystal structures of catalytic RNAs, or fragments thereof, have uncovered novel reoccurring structural elements in RNA and ssDNA.

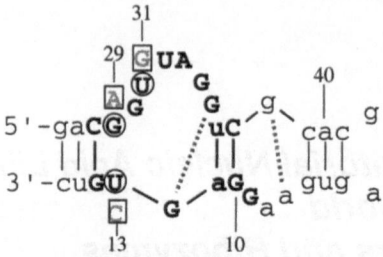

Fig. 1. Secondary structure proposed previously for the citrulline and arginine specific aptamers, based on covariations of selected sequences, on the chemical footprinting pattern obtained in the presence of the cognate amino acid, as well as in damage selection experiments. The bases which were conserved among different isolates are *shown in upper case*, variant bases are *in lower case*. The three nucleotides critical for arginine specificity (13, 29 and 31) are indicated *by circles* (for citrulline) and *boxes* (for arginine)

Between 1978 and 1993 the Brookhaven database contained about 20 high resolution structures of RNA, RNA/DNA complexes, or RNA/protein complexes. In 1996 alone there were no less than 41 structures. The first three-dimensional structures of RNA-aptamer/ligand complexes were reported in 1996 and 1997 – in every case, the structures were resolved by NMR spectroscopy. These studies revealed insight into principles of folding, shape and surfaces, the molecular diversity associated with nucleic acid architecture, molecular recognition and adaptive binding observed in the aptamer/nucleic acid complexes. For example, in the case of the citrulline/arginine aptamer (Burgstaller et al. 1995; Famulok 1994) the three-dimensional structure showed how three mutations within the amino acid binding site of these RNAs determine which of the two amino acids is specifically recognized (Fig. 1; Yang et al. 1996).

Aptamer structures have been comprehensively reviewed and discussed in various commentaries and reviews (Cech and Szewczak 1996; Egli 1997; Feigon et al. 1996; Heus 1997; Patel 1997). Comparisons of various ligand-binding aptamer structures with proteins which bind related molecules showed that nucleic acids and proteins use strikingly similar strategies for the formation of well defined binding pockets. Structures of nucleic acid/ligand complexes that have been published so far are summarized in Table 1.

Table 1. Aptamers for which three-dimensional structural data have been obtained by NMR spectroscopy

Ligand	Binding pocket	Reference
FMN	RNA	Fan et al. 1996
Citrulline	RNA	Yang et al. 1996
Arginine	RNA	Yang et al. 1996
Arginine	DNA	Lin and Patel 1996
AMP	RNA	Dieckmann et al. 1996; Jiang et al. 1996; Nonin et al. 1997
AMP	DNA	Lin and Patel 1997
Tobramycin	RNA	Jiang et al. 1997
Rev peptide	RNA	Ye et al. 1996
Theophylline	RNA	Zimmerman et al. 1997

The molecular discrimination achieved by aptamer/small molecule complexes can be as good as, or even better than, with antibodies. The theophylline aptamer discriminates between the related molecules theophylline and caffeine, which differ by only one methyl group, at least tenfold better than an antibody isolated for the same purpose (Jenison et al. 1994). An aptamer selected for the specific binding of L-arginine shows a 12,000-fold reduced affinity to the D-arginine enantiomer (Geiger et al. 1996). Aptamers for small molecules, such as neomycin (Famulok and Hüttenhofer 1996; Wallis et al. 1995) and flavin mononucleotide (FMN) (Burgstaller and Famulok 1994), have been used in surface plasmon resonance technology to generate target-specific biosensors (Famulok et al., data not shown). By integration of the ATP- or theophylline-binding aptamer sequences into the hammerhead ribozyme (HHR), an allosteric HHR was engineered which carried out the phosphodiester cleavage only after the respective ligands had been added to the cleavage buffer (Tang and Breaker 1997). Another impressive example of how an in vitro selected small-molecule binding RNA can influence cellular processes is a 7-methyl guanosine binding aptamer which inhibits the translation of capped mRNAs but not of uncapped mRNAs in cell-free lysates of HeLa or yeast cells (Haller and Sarnow 1997).

7.2 Functional Aptamers for Proteins and Their Application in Biotechnology, Molecular Medicine, and Diagnostics

The specificity of molecular recognition combined with the ease by which protein-binding aptamers can be isolated, engineered, evolved, and modified chemically – exclusively ex vivo – makes these molecules very attractive as tools in molecular medicine, biotechnology and diagnostics. The international interest in RNA technologies is rapidly increasing. Various recent examples impressively illustrate the wide range of applications aptamers can be used for.

7.2.1 Prion Specific RNA Aptamers

We have recently applied in vitro selection to isolate RNA aptamers which are directed against the Syrian golden hamster cellular prion protein PrP23–231 (PrP^c; Weiss et al. 1997). For the selection a recombinant PrP/glutathione-S-transferase (GST) fusion protein immobilized on glutathione agarose was used (Weiss et al. 1995). Nine iterative cycles of selection and amplification revealed GST::rPrP^C-specific aptamers which did not recognize the fusion partner GST or the fusion protein GST::rPrP90–231 (rPrP27–30) lacking 67 amino acids from the ultimate N-terminus of the prion protein. As this region is also missing in protease K treated PrP^{Sc} isolated from fibrillous plaques of infected species, the aptamers are able to discriminate between PrP^C and PrP27–30. This discriminatory effect does not result from the specific recognition of different three-dimensional structures of the two isoforms PrP^C and PrP^{Sc} thought to be implied in the infectivity of PrP^{Sc}. Instead, the molecular discrimination can be explained on the basis of mapping experiments with GST-PrP peptides which revealed that the region of PrP^c most critical for aptamer binding was the N-terminal amino acids 23–52. Sequence analyses suggest that all aptamers isolated are likely to contain a three-layered G-quartet as a structural element critical for PrP recognition. Individual radiolabeled aptamers specifically recognize authentic prion protein in brain homogenates from various species such as wild-type mice, hamster, and cattle as demonstrated by supershifts obtained in the presence of PrP specific

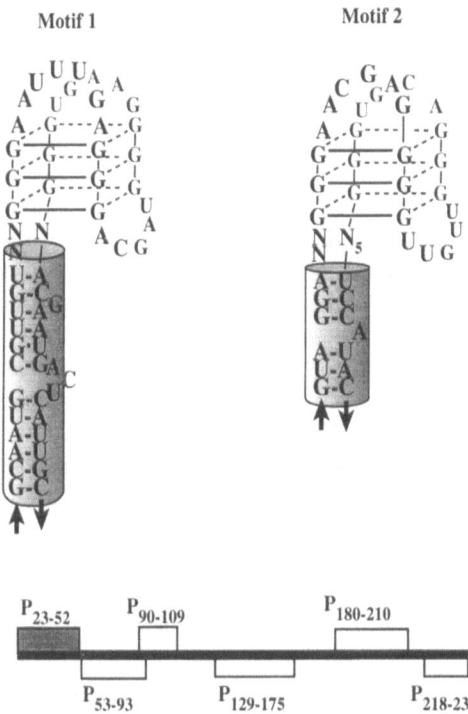

Fig. 2. Secondary structure proposed for two PrP^C-binding aptamers (*Motifs 1 and 2*). Both structures are likely to contain a three-layered G-quartet scaffold, based on sequence conservation. The helical regions shown *within the gray cylinder* can be deleted without significant loss of binding activity. The peptide-constructs used for the mapping of the aptamer binding site on the prion protein are shown *below*. All peptides tested were recombinant PrP-peptide::GST fusion proteins. The fusion peptide corroborating the ultimate N-terminus of the PrP was the only construct which showed a defined and specific gel-shift in the presence of radiolabeled aptamer motifs 1 and 2 in non-denaturing agarose gels

antibodies. No interaction was observed with brain homogenates from PrP knock-out mice (Fig. 2).

The aptamers do not recognize PrP27–30 present in brain homogenates from scrapie infected mice. This study showed that aptamers are able to recognize their specific target within a mixture of hundreds of different proteins contained in tissue homogenates. Aptamers specific for PrP^{Sc} may provide a reliable diagnostic tool that adds up to the long sought and recently obtained PrP^{Sc}-specific monoclonal antibody (Korth et al. 1997). Given the importance of a reliable PrP-diagnostic test, we think that it is generally desirable to have various different PrP^{Sc}-specific molecules available for routine mass PrP^{Sc} diagnostics. A diagnostic nucleic acid, for example, could be used to identify PrP^{Sc} in proteinase treated samples, under which conditions it might be difficult to apply an antibody. When modified nucleic acids are used in the in vitro selection experiment, aptamers can be isolated which are resistant against degradation in biological materials (see below). In addition, aptamers can easily be engineered to meet the criteria required for diagnostic purposes such as stability, size and detectability. We are currently investigating whether our PrP^{C}-specific aptamer is able to reduce or even neutralize infectivity. Such an activity might be of potential therapeutic interest.

We also looked at another direction when we applied the aptamer technology to the prion problem: a prion-specific aptamer might provide useful to develop an assay for the PrP$^{C}{\rightarrow}$PrP^{Sc} transition. We hypothesized that an aptamer might have a sort of chaperone-like function in this transition and, finally, that such an aptamer might reflect the function of a possible natural partner as predicted by the co-prion hypothesis. We shall see!

7.2.2 Functional Nuclease Resistant Aptamers

The PrP-binding aptamers exploited sufficient stability in crude brain homogenates probably due to protection from exonuclease degradation by the stable G-quartet scaffold (Fig. 2a). Small protein binding RNA aptamers, however, usually do not contain stabilizing structural scaffolds of that kind. This has led to a widely held bias against nucleic acids as potential therapeutics, diagnostics, or assay components. However, a number of recent studies have demonstrated that functional nucleic acids not only can be made strikingly small, but also resistant against degradation in biological materials (Eaton 1997; Klußmann et al. 1996; Nolte et al. 1996; Osborne and Ellington 1997). One way to circumvent the vulnerability of RNA to nuclease degradation is indirect: in a first step an aptamer that binds the enantiomer of the target is selected, then, in a second step, the enantiomer of the aptamer is synthesized (from L-phosphoramidites) as a nuclease-insensitive ligand of the natural target. This mirror-image approach, sometimes also designated as the "Spiegelmer" approach, has been applied to L-arginine (Nolte et al. 1996), D-adenosine (Klußmann et al. 1996), and the peptide hormone vasopressin (Williams et al. 1997).

An alternative approach is the direct selection of an aptamer from libraries of modified RNAs. Modifications have to be chosen so as to be compatible with nucleic acid replicating enzymes such as reverse transcriptase, or DNA- and RNA-polymerases. The modifications most commonly used are those in which the 2'-OH group of pyrimidines is substituted by a 2'-fluoro-, or a 2'-amino group (Aurup et al. 1994; Pieken et al. 1991). 2'-Amino-modified nuclease resistant aptamers were selected which bind to autoantibodies of patients affected by the muscular disease *myasthenia gravis*. These aptamers inhibit the binding of acetylcholine receptors from human cells to these autoantibodies and block the pathogenic consequences associated with this (Lee and Sullenger 1997). Similarly, 2'-fluoro-modified nuclease resistant aptamers directed against the human keratinocyte growth factor block its activity with a K_i of 34 pM (Pagratis et al. 1997).

An aptamer with 2'-amino pyrimidine modifications selected for binding to vascular permeability factor/vascular endothelial growth factor (VPF/VEGF) was minimized to a modified 24-mer. The 2'-OH groups of defined purine residues were subsequently modified by 2'-methoxy groups in a damage selection experiment in which variants of this aptamer transcribed with a mixture of 2'-OH-purines and 2'-OCH$_3$-purines were screened for enhanced binding to VPF/VEGF (Green et al. 1995). 2'-Methoxy substituted purine residues in this aptamer were identified by protection from alkaline hydrolysis. Among the 13 purines in the 24-mer, 9 could be substituted by the 2'-methoxy group; the other 4 could not be changed without significant loss of binding affinity (Fig. 3).

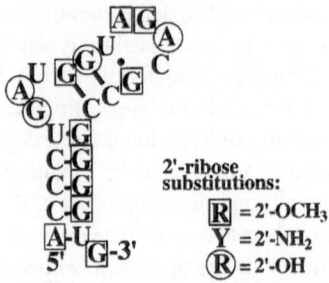

Fig. 3. Secondary structure of the 24-mer VPF/VEGF modified aptamer. Pyrimidines (*Y*) all contained 2'-fluoro-2'-deoxy modifications, purines (*R*) with 2'-methoxy-2'-deoxy modifications are *boxed*, purines (*R*) with unmodified ribose sugar residues are *circled*. The dissociation constant of the aptamer/protein complex was $0.19 \, \text{n}M$

The final modified aptamer bound to the target protein with a K_d of $0.14 \, \text{n}M$ and specifically blocked the binding of ^{123}I-labeled VPF/VEGF to cell surface receptors expressed on human umbilical vein endothelial cells.

7.2.3 Functional Aptamers In Vivo

For the first time some RNA aptamers that were isolated in vitro have also been expressed in vivo to study their biological function within a cell. Among them are an anti-HIV-1 Rev aptamer (Good et al. 1997) which inhibits HIV-1-infection in the cells in which it is expressed, a series of aptamers which bind the special prokaryotic elongation factor SelB (Conrad et al. 1997; Klug et al. 1997), and an aptamer specific for RNA polymerase II from yeast which was shown to inhibit yeast growth specifically (Thomas et al. 1997).

One of the most impressive recent examples for aptamer application is the use of a modified DNA aptamer inhibitor for human neutrophil elastase, a serine protease, for the in vivo diagnostic imaging of inflammation (Charlton et al. 1997b). The aptamer was isolated by a technique called "blended SELEX," in which a weakly reactive elastase inhibitor was incorporated into all pool members of a randomized DNA library

(Charlton et al. 1997a). The aptamers that best promoted the covalent reaction of the reactive moiety inactivated elastase with a k_{obs} around $2 \times 10^8 M^{-1}$ min^{-1}, almost two orders of magnitude faster than peptide-based inhibitors. In a parallel study, the aptamer inhibitor was also shown to reduce lung injury in a rat alveolitis model (Bless et al. 1997). By using the ^{99m}Tc-labeled anti elastase aptamer and, as a control, a ^{99m}Tc-labeled rat anti-elastase IgG which is clinically used in inflammatory in vivo imaging, it was shown that the aptamer achieved a significantly higher target-to-background (T/B) ratio in less time than the IgG. This example indicates that in some cases specific ligand binding nucleic acids might be advantageous over antibodies; in the present case the superior T/B ratios were attributed to the more rapid clearance of the aptamer from the peripheral circulation compared to the IgG.

7.3 Catalytic Nucleic Acids
Selected from Combinatorial Libraries

In vitro selection was also used to isolate a considerable number of novel ribozymes which impressively enlarge the scope of RNA catalyzed chemical transformations. The latest results in this field demonstrate that ribozymes or deoxyribozymes are able to catalyze a wide range of chemical reactions. They are summarized in various comprehensive reviews (Breaker 1997a–c; Burgstaller and Famulok 1998; Jaeger 1997; Narlikar and Herschlag 1997; Pan 1997). The most recent advances include ribozymes for: the Diels-Alder reaction (Tarasow et al. 1997), amide bond formation (Lohse and Szostak 1996; Wiegand et al. 1997), peptide bond formation (Zhang and Cech 1997), a ribozyme with novel oligonucleotide ligation activity (Hager and Szostak 1997), and a ribozyme for the formation of 5'-5'-diphoshate bonds (Huang and Yarus 1997). Of great importance for evolutionary RNA biotechnology and molecular evolution is a novel technique which allows the continuous evolution of ribozymes (Wright and Joyce 1997). This process allows for the amplification, mutation, and selection of many ribozyme generations within a very short period of time.

We have recently isolated a new ribozyme which catalyzes the transfer of an amino acid ester from a biotinyl-N-phenylalanyl-2'(3')-adeno-

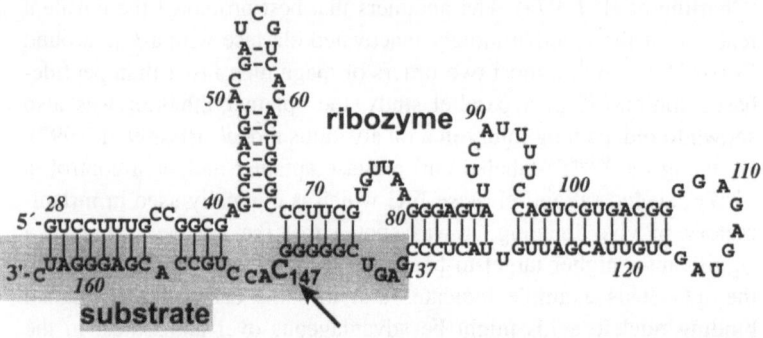

Fig. 4. Secondary structure of the clone 11 transacylase ribozyme based on the Zuker RNA folding algorithm Mfold. The oligonucleotide substrate is *shaded in gray*. The 2'-OH group of cytosine 147 (*arrow*) is the site of modification of the oligonucleotide substrate (Jenne and Famulok 1998)

sine-5'-monophosphate (Bio-Phe-AMP) substrate onto a specific ribose 2'-OH group (Fig. 4; Jenne and Famulok 1998).

This ester transferase ribozyme is thus an example of an RNA which catalyzes a reaction at a carbon center by utilizing a low molecular weight cofactor. The reaction strongly depends on the presence of divalent metal ions and can be inhibited by AMP, but not with GMP, indicating that a specific binding pocket for the Bio-Phe-AMP substrate exists. The transformation reaches equilibrium due to a significant level of the corresponding reverse reaction, 2'(3')-aminoacylation of AMP. The reverse reaction was used to determine the equilibrium constant of the ester transfer reaction which lies strongly on the side of the 2'(3')-aminoacyl AMP educt. The central role aminoacylated RNAs play in translation processes suggests that acyl transfer reactions catalyzed by RNA might have facilitated the development and optimization of the translation apparatus during early evolution. It has been proposed that the evolutionary precursor of rRNA might have been a ribozyme with 3'-OH→2'(3')-OH acyl-transferase activity (Hager et al. 1996). Our ribozyme expands the scope of RNA catalysis towards this direction and shows that such transformations are – in principle – possible.

Despite the general acceptance that the 2'-hydroxyl group of RNA is essential for its capability to form complex secondary and tertiary structures and therefore for its catalytic activity, novel results of in vitro selection experiments aimed towards the isolation of deoxyribozymes make clear that DNA also can adopt a variety of complex structures enabling it to catalyze a wide range of chemical reactions (Breaker 1997a–c). The man-made ribo- and deoxyribozymes will certainly be applicable in some diagnostic, technical or medical purposes, where it is advantageous that they are based not on proteins but on nucleic acids.

We expect that we will soon see examples of ribo- and deoxyribozymes evolved for the catalysis of complex chemical transformations. There is enough reason to assume that such synthetic enzymes will be used as catalysts in organic syntheses. The novel catalysts not only support theories of an "RNA world," in which the metabolism and replication of primitive organisms were controlled by RNA enzymes (Hager et al. 1996). There is also the potential for biotechnological, synthetic, or diagnostic applications.

References

Aurup H, Tuschl T, Benseler F, Ludwig J, Eckstein F (1994) Oligonucleotide duplexes containing 2'-amino-2'-deoxycytidines: thermal stability and chemical reactivity. Nucleic Acids Res 22:20–24

Bless NM, Smith D, Charlton J, Czermak BJ, Schmal H, Friedl HP, Ward PA (1997) Protective effects of an aptamer inhibitor of neutrophil elastase in lung inflammatory injury. Curr Biol 7:877–880

Breaker RR (1997a) DNA aptamers and DNA enzymes. Curr Opin Chem Biol 1:26–31

Breaker RR (1997b) DNA enzymes. Nature Biotechnol 15:427–431

Breaker RR (1997c) In vitro selection of catalytic polynucleotides. Chem Rev 97:371–390

Burgstaller P, Famulok M (1994) Isolation of RNA aptamers for biological co-factors by in-vitro selection. Angew Chem Int Ed Engl 33:1084–1087

Burgstaller P, Famulok M (1998) Synthetic ribozymes and deoxyribozymes. In: Waldmann H, Mulzer J (eds) Organic synthesis highlights, vol 3. Wiley-VCH, Weinheim (in press)

Burgstaller P, Kochoyan M, Famulok M (1995) Structural probing and damage selection of citrulline and arginine-specific RNA aptamers identify base positions required for binding. Nucleic Acids Res 23:4769–4776

Cech TR, Szewczak AA (1996) Selecting apt RNAs for NMR. RNA 2:625–627

Charlton J, Kirschenheuter GP, Smith D (1997a) Highly potent irreversible inhibitors of neutrophil elastase generated by selection from a randomized DNA-valine phosphonate library. Biochemistry 36:3018–3026

Charlton J, Sennello J, Smith D (1997b) In vivo imaging of inflammation using an aptamer inhibitor of human neutrophil elastase. Chem Biol 4:809–816

Conrad RC, Symensma TL, Ellington AD (1997) Natural and unnatural answers to evolutionary questions. Proc Natl Acad Sci USA 94:7126–7128

Dieckmann T, Suzuki E, Nakamura GK, Feigon J (1996) Solution structure of an ATP-binding RNA aptamer reveals a novel fold. RNA 2:628–640

Eaton BE (1997) The joys of in vitro selection: chemically dressing oligonucleotides to satiate protein targets. Curr Opin Chem Biol 1:10–16

Egli M (1997) Wie in vitro selektierte RNA-Aptamere funktionieren-die räumlichen Strukturen von Substrat-RNA-Aptamer-Komplexen. Angew Chem 109:494–497

Famulok M (1994) Molecular recognition of amino acids by RNA-aptamers: an l-citrulline binding RNA motif and its evolution into an l-arginine binder. J Am Chem Soc 116:1698–1706

Famulok M, Hüttenhofer A (1996) In vitro selection analysis of neomycin binding RNAs with a mutagenized pool of variants of the 16 S rRNA decoding region. Biochemistry 35:4265–4270

Fan P, Suri AK, Fiala R, Live D, Patel DJ (1996) Molecular recognition in the FMN-RNA aptamer complex. J Mol Biol 258:480–500

Feigon J, Dieckmann T, Smith FW (1996) Aptamer structures from A to ζ. Chem Biol 3:611–617

Geiger A, Burgstaller P, von der Eltz H, Roeder A, Famulok M (1996) RNA aptamers that bind L-arginine with sub-micromolar dissociation constants and high enantioselectivity. Nucleic Acids Res 24:1029–1036

Good PD, Krikos AJ, Li SXL, Bertrand E, Lee NS, Giver L, Ellington A, Zaia JA, Rossi JJ, Engelke DR (1997) Expression of small, therapeutic RNAs in human cell nuclei. Gene Ther 4:45–54

Green LS, Jellinek D, Bell C, Beebe LA, Feistner BD, Gill SC, Jucker FM, Janjic N (1995) Nuclease-resistant nucleic acid ligands to vascular permeability factor/vascular endothelial growth factor. Chem Biol 2:683–695

Hager AJ, Szostak JW (1997) Isolation of novel ribozymes that ligate AMP-activated RNA substrates. Chem Biol 4:607–617

Hager AJ, Pollard JD, Szostak JW (1996) Ribozymes: aiming at RNA replication and protein synthesis. Chem Biol 3:717–725

Haller AA, Sarnow P (1997) In vitro selection of a 7-methyl-guanosine binding RNA that inhibits translation of capped mRNA molecules. Proc Natl Acad Sci USA 94:8521–8526

Heus HA (1997) RNA aptamers. Nat Struct Biol 4:597–600

Huang F, Yarus M (1997) Versatile 5' phosphoryl coupling of small and large molecules to an RNA. Proc Natl Acad Sci USA 94:8965–8969

Jaeger L (1997) The new world of ribozymes. Curr Opin Struct Biol 7:324–335

Jenison RD, Gill SC, Pardi A, Polisky B (1994) High-resolution molecular discrimination by RNA. Science 263:1425–1429

Jenne A, Famulok M (1998) A novel ribozyme with ester transferase activity. Chem Biol 5:23–34

Jiang F, Kumar RA, Jones RA, Patel DJ (1996) Structural basis of RNA folding and recognition in an AMP-RNA aptamer complex. Nature 382:183–186

Jiang L, Suri AK, Fiala R, Patel DJ (1997) Saccharide-RNA recognition in an aminoglycoside antibiotic-RNA aptamer complex. Chem Biol 4:35–50

Klug SJ, Hüttenhofer A, Kromayer M, Famulok M (1997) In vitro and in vivo characterization of novel mRNA motifs that bind special elongation factor SelB. Proc Natl Acad Sci USA 94:6676–6681

Klußmann S, Nolte A, Bald R, Erdmann VA, Fürste JP (1996) Mirror-image RNA that binds D-adenosine. Nat Biotechnol 14:1112–1115

Korth C, Stierli B, Streit P, Moser M, Schaller O, Fischer R, Schulz-Schaeffer W, Kretzschmar H, Raeber A, Braun U, Ehrensperger F, Hornemann S, Glockshuber R, Riek R, Billeter M, Wüthrich K, Oesch B (1997) Prion (PrPSc)-specific epitope defined by a monoclonal antibody. Nature 390:74–77

Lee S-W, Sullenger BA (1997) Isolation of a nuclease-resistant decoy RNA that can protect human acetylcholine receptors from myasthenic antibodies. Nature Biotechnology 15:41–45

Lin CH, Patel DJ (1996) Encapsulating an amino acid in a DNA fold. Nat Struct Biol 3:1046–1050

Lin CH, Patel DJ (1997) Structural basis of DNA folding and recognition in an AMP–DNA aptamer complex: distinct architectures but common recognition motifs for DNA and RNA aptamers complexed to AMP. Chem Biol 4:817–832

Lohse PA, Szostak JW (1996) Ribozyme-catalysed amino acid transfer reactions. Nature 381:442–444

Narlikar GJ, Herschlag D (1997) Mechanistic aspects of enzymatic catalysis: lessons from comparison of RNA and protein enzymes. Annu Rev Biochem 66:19–59

Nolte A, Klußmann S, Bald R, Erdmann VA, Fürste JP (1996) Mirror-design of L-oligonucleotide ligands binding to L-arginine. Nat Biotechnol 14:1116–1119

Nonin S, Jiang F, Patel DJ (1997) Imino proton exchange and base-pair kinetics in the AMP-RNA aptamer complex. J Mol Biol 268:359–374

Osborne SE, Ellington AD (1997) Nucleic acid selection and the challenge of combinatorial chemistry. Chem Rev 97:349–370

Pagratis NC, Bell C, Chang Y-F, Jennings S, Fitzwater T, Jellinek D, Dang C (1997) Potent 2'-amino-, and 2'-fluoro-2'-deoxyribonucleotide RNA inhibitors of keratinocyte growth factor. Nature Biotechnology 15:68–73

Pan T (1997) Novel and variant ribozymes obtained through in vitro selection. Curr Opin Chem Biol 1:17–25

Patel D (1997) Structural analysis of nucleic acid aptamers. Curr Opin Chem Biol 1:32–46

Pieken WA, Olsen DB, Benseler F, Aurup H, Eckstein F (1991) Kinetic characterization of ribonuclease-resistant 2'-modified hammerhead ribozymes. Science 253:314–317

Tang J, Breaker RR (1997) Rational design of allosteric ribozymes. Chem Biol 4:453–459

Tarasow TM, Tarasow SL, Eaton BE (1997) RNA-catalysed carbon-carbon bond formation. Nature 389:54–57

Thomas M, Chédin S, Carles C, Riva M, Famulok M, Sentenac A (1997) Selective targeting and inhibition of yeast RNA polymerase II by RNA aptamers. J Biol Chem 272:27980–27986

Wallis MG, Von Ahsen U, Schroeder R, Famulok M (1995) A novel RNA motif for neomycin recognition. Chem Biol 2:543–552

Weiss S, Proske D, Neumann M, Groschup MH, Kretzschmar HA, Famulok M, Winnacker E-L (1997) RNA aptamers specifically interact with the prion protein PrP. J Virol 71:8790–8797

Weiss S, Famulok M, Edenhofer F, Wang Y-H, Jones IM, Groschup M, Winnacker E-L (1995) Overexpression of Active Syrian Golden hamster prion protein PrP^c as a glutathione S-transferase fusion in heterologous systems. J Virol 69:4776–4783

Wiegand TW, Janssen RC, Eaton BE (1997) Selection of RNA amide synthases. Chem Biol 4:675–683

Williams KP, Liu X-H, Schumacher TNM, Lin HY, Ausiello DA, Kim PS, Bartel DP (1997) Bioactive and nuclease-resistant L-DNA ligand of vasopressin. Proc Natl Acad Sci USA 94:11285–11290

Wright MC, Joyce GF (1997) Continuous in vitro evolution of catalytic function. Science 276:614–617

Yang Y, Kochoyan M, Burgstaller P, Westhof E, Famulok M (1996) Structural basis of ligand discrimination by two related RNA aptamers resolved by NMR spectroscopy. Science 272:1343–1347

Ye X, Gorin A, Ellington AD, Patel DJ (1996) Deep penetration of an α-helix into a widened RNA major groove in the HIV-1 rev peptide-RNA aptamer complex. Nat Struct Biol 3:1026–1033

Zhang B, Cech TR (1997) Peptide bond formation by in vitro selected ribozymes. Nature 390:96–100
Zimmerman GR, Jenison RD, Wick CL, Simorre J-P, Pardy A (1997) Inter-
   locking structural motifs mediate molecular discrimination by a theophylline-binding RNA. Nat Struct Biol 4:644–649

8 Sequence Specific Recognition of Double Stranded DNA by Peptide Nucleic Acid

P.E. Nielsen

8.1   Introduction 151
8.2   Peptide Nucleic Acid 155
8.3   DNA Targeting by PNA 157
8.3.1   Triplex Binding 157
8.3.2   Triplex Invasion 159
8.3.3   Duplex Invasion 162
8.3.4   Double Duplex Invasion 163
8.4   Concluding Remarks 164
References 164

8.1 Introduction

Sequence selective recognition of double stranded DNA by proteins is crucial for cellular control and regulation of gene expression, and evolution has selected several protein structural motifs, including α-helices, β-strands and Zn-fingers, to accomplish this (e.g., Rhodes et al. 1996; Nelson 1995).

The DNA double helix contains "blueprints" of the base pairs that are accessible from the outside both from the minor and especially from the major groove (Fig. 1). By molecular complementarity these hydrogen bond donor and acceptor sites as well as the hydrophobic methyl groups of thymine in combination with the overall shape of the groove

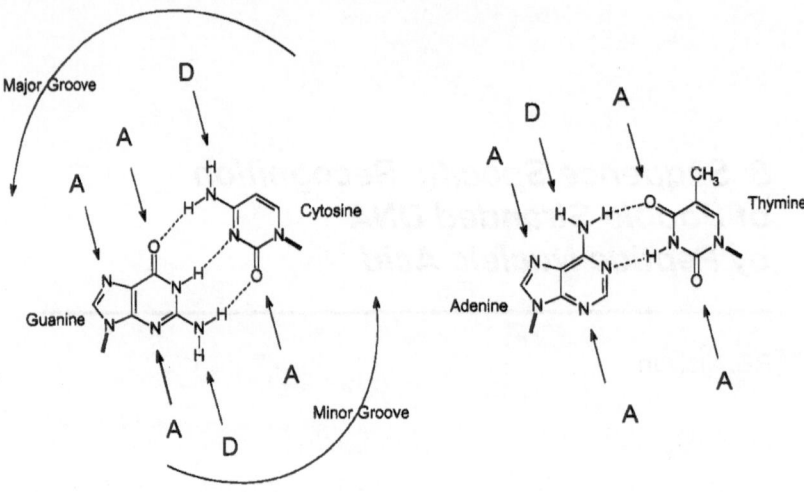

Fig. 1. Base pair recognition. Hydrogen bonding donor (*D*) and acceptor (*A*) patterns of DNA base pairs in the major and minor grooves of a double helix

are used by the proteins for recognition, and Nature seems almost exclusively to have chosen the major groove, which also contains the greater amount of information. Electrostatic interactions primarily with the phosphate backbone supply additional binding energy.

Unfortunately, the DNA recognition by proteins is best described as analogous, i.e., by shape/molecular complementarity, rather than digital, i.e., direct reading of consecutive base pairs, and no base pair-(oligo)peptide code appears deducible for any of the three protein recognition motifs (α-helix, β-sheet, Zn-finger), and only very recently using the empirical "rules" of a Zn-finger-3-base pair "recognition code" (Choo and Klug 1995, 1997) has it now become possible to design proteins targeted to the desired sequence.

Instead the chemical research on design of sequence specific DNA binding ligands has mostly concentrated on minor groove binders inspired by the natural drugs distamycin and netropsin and major groove binding by oligonucleotides. More recently, the double helix invading peptide nucleic acids (PNAs) have entered the scene (Nielsen 1997).

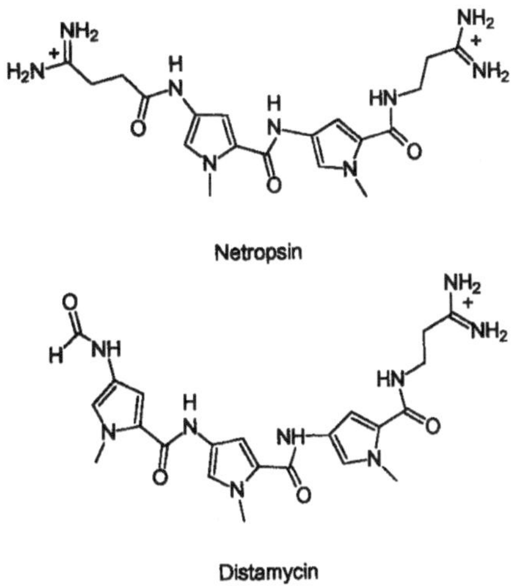

Fig. 2. Chemical structures of the DNA minor groove binders netropsin and distamycin

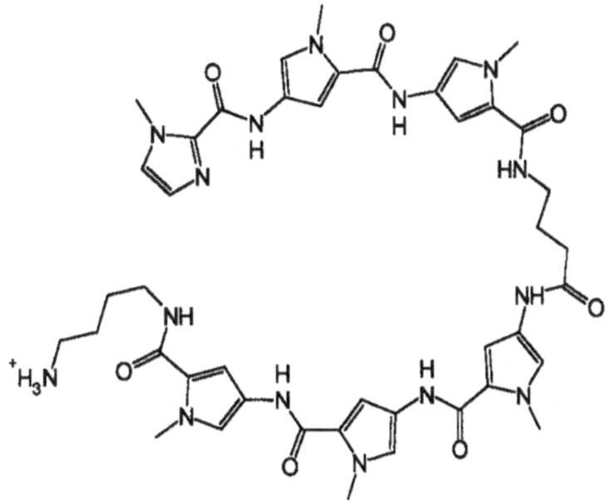

Fig. 3. Chemical structure of the hairpin polyamide, where a pyrrole unit recognizes half of an A-T base pair or the cytosine of a C-G base pair, whereas an imidazole recognizes the guanine of a G-C base pair

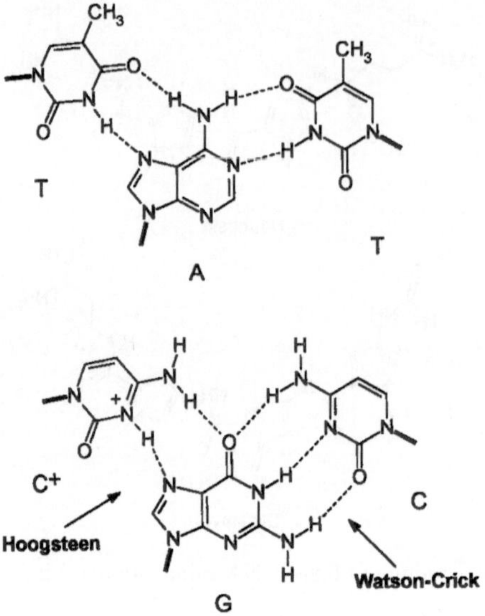

Fig. 4. Recognition of A-T and G-C base pairs by thymine and cytosine Hoogsteen base pairing. Note that the cytosine requires protonation on N3

The natural minor groove binders (such as distamycin and netropsin) (Fig. 2) exhibit very high specificity of $(A/T)_n$ ($n \geq 4$) regions of the DNA, and earlier attempts to modify these ligands also to allow targeting of G/C containing sequences were met with only very modest success (Lown 1994). However, a seminal discovery in 1989 by Pelton and Wemmer showing by NMR spectroscopy that two molecules of distamycin may bind side by side in the minor groove of the DNA helix opened a novel avenue for minor groove DNA recognition. Subsequent elaborate and elegant work by the Dervan group (e.g., Wade et al. 1992; Parks et al. 1996; Swalley et al. 1997) has now resulted in a set of recognition rules for hair-pin minor groove binders constructed from N-Me-pyrrole and imidazole units (Fig. 3) that allow design of very tight binding ligands for shorter mixed sequence duplex DNA targets (White et al. 1997). Although the general validity as well as the exact specificity of these rules is not yet firmly established, this appears a very promising approach to sequence specific DNA targeting.

Triplex forming polynucleotides were discovered more than 40 years ago (Felsenfeld et al. 1957), but only 10 years ago (LeDoan et al. 1987; Moser and Dervan 1987) was it realized and demonstrated that also shorter oligonucleotides (~15 nt) could bind to a duplex target and thus possibly be developed into sequence targeted DNA ligands (Dervan 1992; Hélène 1993). However, due to the Hoogsteen recognition mode of triplex formation (Fig. 4), this recognition was confined to homopurine targets. Despite many efforts, including alternative triplet motifs, use of alternative (non-natural) nucleobases and strand switching strategies, oligonucleotide triplex targeting of duplex DNA is essentially still restricted to homopurine (rich) sequences. However, triplex targeting is inherently digital, a feature that makes this binding mode very attractive for rational design.

8.2 Peptide Nucleic Acid

Peptide nucleic acid (PNA; Nielsen et al. 1991; Egholm et al. 1992; Fig. 5) was originally designed as a DNA mimic for triplex targeting, but most surprisingly subsequent experiments revealed that ho-

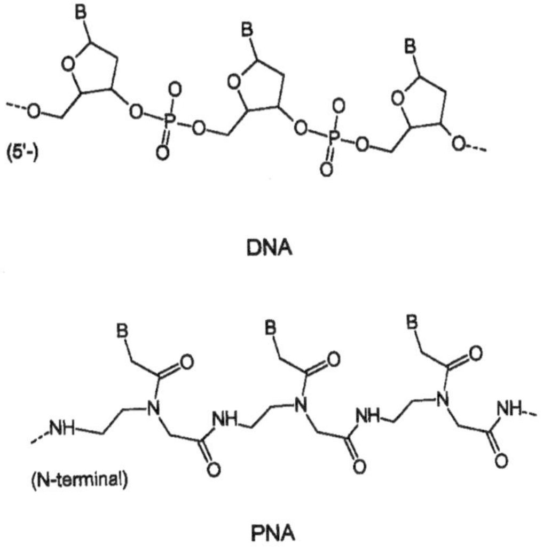

Fig. 5. Chemical structures of DNA and PNA

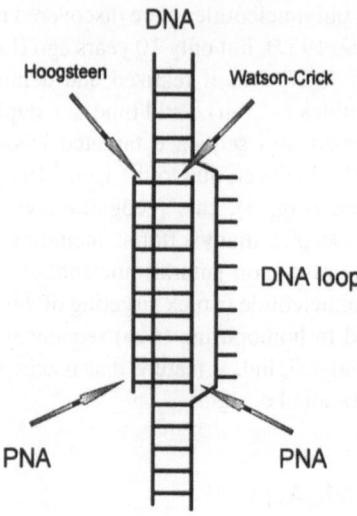

Fig. 6. Schematic drawing of a PNA$_2$-DNA triplex strand displacement complex, showing the two PNA oligomers forming Watson-Crick and Hoogsteen base pairs with the complementary DNA strand, while the noncomplementary DNA strand is left as a single stranded loop

mopyrimidine PNAs prefer binding their homopurine targets in duplex DNA by strand displacement involving an internal PNA$_2$-DNA triplex of extraordinary stability (Nielsen et al. 1991, 1994; Cherny et al. 1993; Fig. 6). This novel binding mode naturally attracted much attention and provided new opportunities for developing sequence specific duplex DNA targeted ligands which warranted further investigation and development (Hyrup and Nielsen 1996; Good and Nielsen 1997).

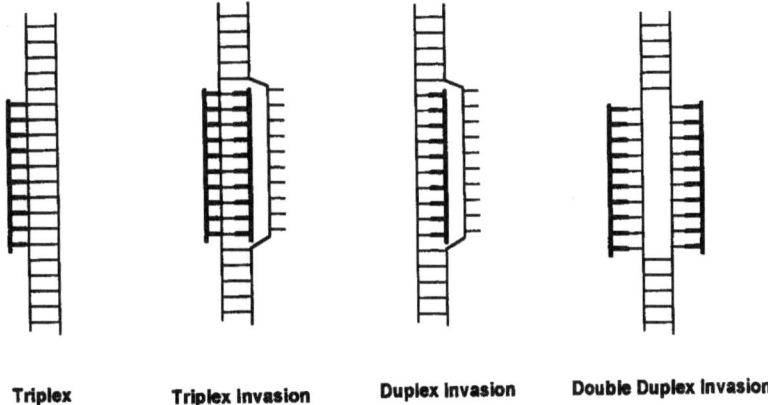

Triplex   **Triplex Invasion**   **Duplex Invasion**   **Double Duplex Invasion**

Fig. 7. Schematic drawing of the various modes by which PNA oligomers have been found to recognize double stranded DNA

8.3 DNA Targeting by PNA

8.3.1 Triplex Binding

Since the introduction of PNA more than 6 years ago, many aspects of the interaction of these oligomers with double stranded DNA have been investigated and clarified. In fact, it has recently been shown that some homopyrimidine PNA oligomers may indeed recognize double stranded DNA by external triple helix binding in the major groove (Fig. 7) as originally designed. However, such PNA-DNA$_2$ triplexes appear much less stable than the corresponding PNA$_2$-DNA-DNA triplex strand displacement complexes. Also they seem to be only observed with cytosine-rich PNAs (Praseuth et al. 1996; Wittung et al. 1997), and although they could be intermediates towards the more stable triplex invasion complex (Fig. 8), the available evidence therefore argues against such a mechanism.

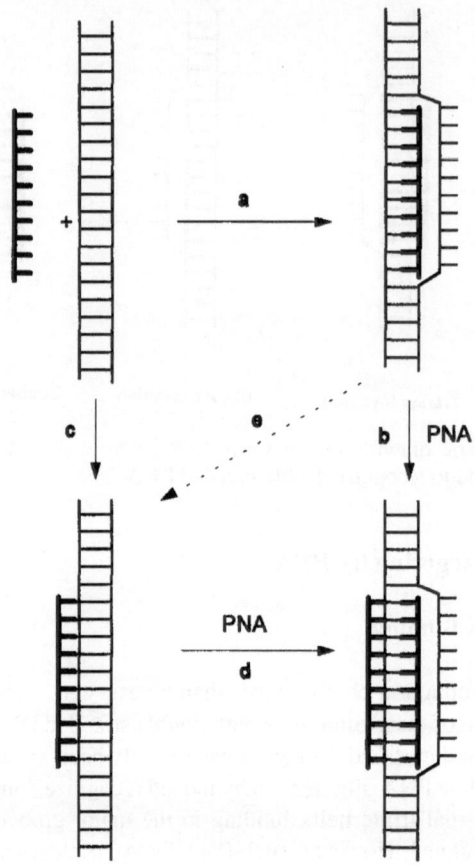

Fig. 8. Schematic drawing of the simplest possible pathways to a duplex strand displacement complex

8.3.2 Triplex Invasion

PNA triplex strand displacement complexes are exceedingly stable. For instance, it has been found that the complex between PNA T_{10} and a double stranded A_{10} target has a thermal stability above 70°C (Nielsen et al. 1993), has a half-life counted in weeks, and is resistant to an ionic strength of at least 500 mM (Cherny et al. 1993). Nonetheless, the binding shows a surprisingly high sequence discrimination (more than 100-fold against a single G-mismatch in the middle of the target), which was explained by the finding that discrimination is almost exclusively kinetically controlled (Demidov et al. 1993; Kuhn et al. 1998).

Since the invasion process requires the DNA double helix to open at least transiently, it is not surprising that the on-rate for strand displacement binding is comparably small and exhibits a high sensitivity to the ionic strength of the medium (Peffer et al. 1993; Demidov et al. 1995). In fact, binding of simple PNAs to their double stranded DNA targets at physiological ionic strength (e.g., 140 mM K^+) cannot be detected even at high micromolar concentrations of PNA and 24 h incubation. Furthermore, triplex invasion by PNA is subject to similar sequence restrictions that govern the triplex formation by DNA oligonucleotides. Thus efficient binding is only observed with homopurine targets (and therefore homopyridimide PNAs), and if the PNA contains cytosines – which require N3-protonation for Hoogsteen binding to guanine – the binding is only efficient at acidic pH (pH≤6).

Fortunately, some rather simple chemical modifications of the PNA oligomers can improve their binding characteristics significantly. Since two PNA oligomers are required for triplex invasion and these, furthermore, should be antiparallel to each other (antiparallel binding is preferred for Watson-Crick binding, whereas the parallel configuration is most favorable for the Hoogsteen strand), it is rather straightforward to connect the two PNAs into a bis-PNA by continuous synthesis (Egholm et al. 1995). This also gives the freedom to choose nucleobases optimal for Watson-Crick and Hoogsteen interactions independently. Thus exchange of cytosine for pseudoisocytosine (Fig. 9) in the Hoogsteen strand of a bis-PNA effectively eliminates the pH sensitivity of the binding (Egholm et al. 1995). Simply, making cationic PNAs by supplying extra lysine residues improves their binding rate significantly (Fig. 10), most likely by ensuring a high local concentration of the PNA

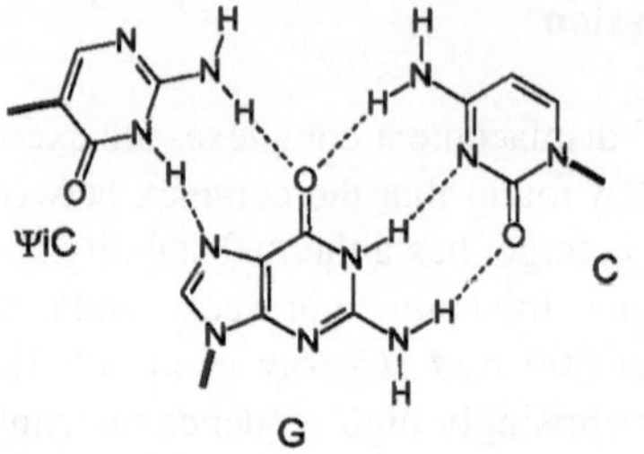

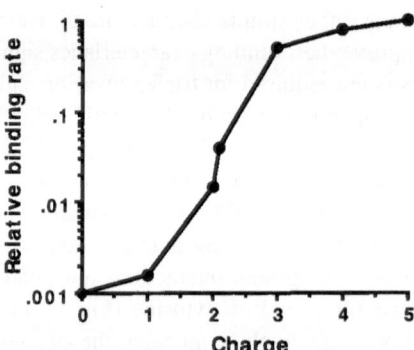

Fig. 9. Chemical structure of pseudoisocytosine (ΨiC) and the recognition of the G-C base pair

at the DNA helix, thereby increasing the probability of the PNA "catching" a transiently open DNA region formed via natural dynamic "breathing" of the DNA double helix (Bentin and Nielsen 1996). Indeed, by combining these three features, it is possible to obtain bis-PNAs which bind their double stranded DNA target so efficiently that binding even at 140 mM K$^+$ is possible with low micromolar concentrations of the PNA (Fig. 11; Griffith et al. 1995). Most importantly this increase in binding rate is accomplished without any significant sacrifice in sequence discrimination (Kuhn et al. 1998).

Much effort has during the last 10 years been invested in finding or developing nucleobases which in a Hoogsteen mode would recognize

Fig. 10. Effect of the number of positive charges (lysines) of a PNA decamer on the relative binding rate of this PNA to its complementary target in double stranded DNA (Nielsen and Demidov, unpublished)

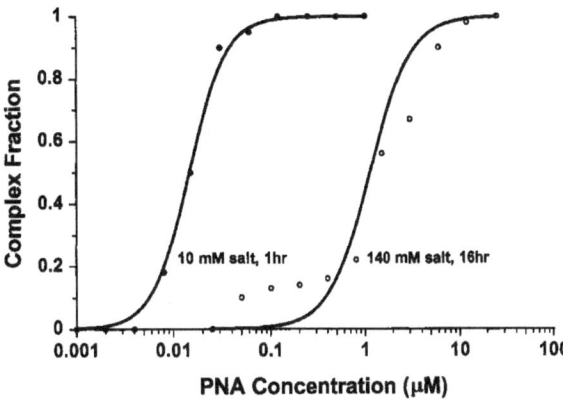

Fig. 11. Binding isotherms of the binding of a bis-PNA (H-Lys-Lys-TTJTTJTTTT-(eg1)$_3$-TTTTCTTCTT-LysNH$_2$) to its complementary target in double stranded DNA at 10 mM K$^+$, pH 7 (1 h incubation time), or at 140 mM K$^+$, pH 7 (16 h incubation time; Nielsen, unpublished)

cytosine or thymine, thereby completing a full four-letter "Hoogsteen alphabet" and allow for general sequence recognition via triplex binding. PNAs could significantly reduce the obstacles towards this goal, both due to the more versatile chemistry in preparing monomers, but also due to the fact that half of the recognition is already accomplished via the Watson-Crick strand and therefore nucleobases that merely tolerate cytosine and thymine rather than specifically recognize these could be sufficient.

In the first step towards a nucleobase that would recognize thymine, we recently developed the pyridazinone "E-base" (Eldrup et al. 1997; Fig. 12). Thymine recognition represents a special problem due to the steric interference of the methyl-group. In order to address this, we used computer modeling, and our first generation ligand – which does indeed recognize thymine although with less than desired affinity – was built with an extended linker to circumvent the thymine methyl group and a tertiary nitrogen in the 2-position to avoid a hydrogen that would also clash with the methyl group (Fig. 12).

Although the efforts by us and others (see references in Eldrup et al. 1997) in the development of novel nucleobases have so far not yet led to a general triplex recognition principle, they do hold some promise that

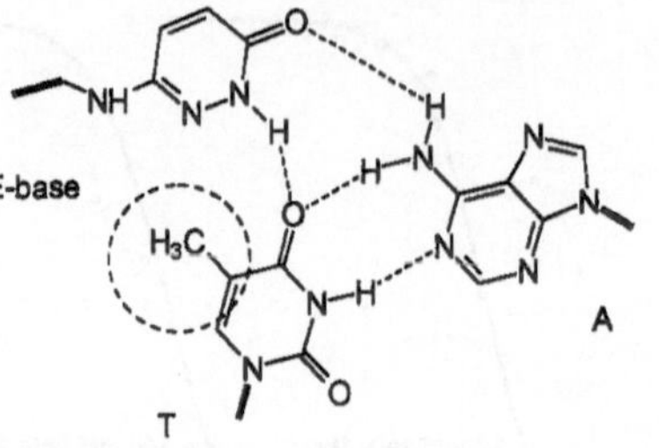

Fig. 12. Chemical structure of the "E-base" and the proposed mode of recognition of the T-A base pair. The steric interference of the methyl group of thymine is indicated

thymine (and cytosine) recognizing ligands can be developed not only in a PNA, but also in a DNA context.

8.3.3 Duplex Invasion

The necessity of developing a new triplex nucleobase would be totally redundant if strand displacement could be accomplished without the requirement of the second PNA strand, i.e, by using only duplex forming PNAs, since these would rely solely on Watson-Crick base pairing. Unfortunately, normal mixed purine-pyrimidine sequence PNAs do not generally bind to double stranded DNA targets. It has been found, however, that a homopurine PNA (H-AAAAGGAGAG-Lys-NH$_2$) that forms only a duplex with its complementary DNA does indeed bind its double stranded DNA target by strand displacement (Nielsen and Christensen 1996). This seems possible because of the extraordinary high thermal stability (Tm=70°C) of the duplex between this homopurine PNA and the complementary DNA.

These results demonstrate that it may indeed be possible to efficiently target double stranded DNA by simple duplex invasion via Watson-Crick recognition provided nucleobase or backbone modified PNAs (or other DNA mimics) with greatly improved binding affinities can be devised.

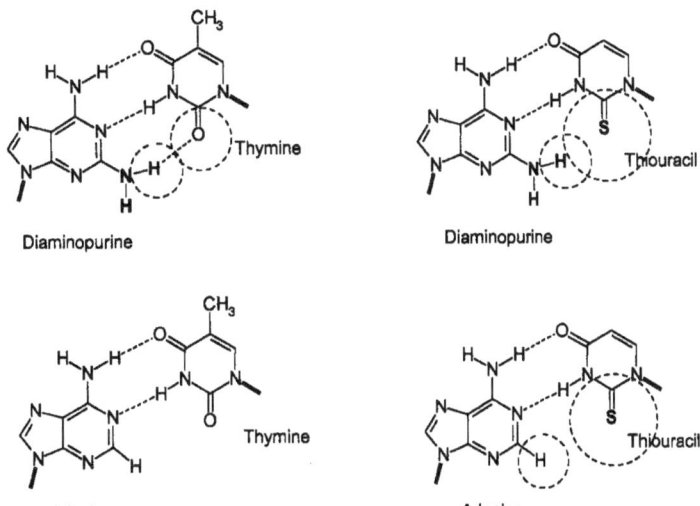

Fig. 13. Chemical structures of diaminopurine-thymine, diaminopurine-thiouracil, adenine-thymine and adenine-thiouracil base pairs, highlighting the extra hydrogen bond formed by the 2-amino group of diaminopurine with the 2-oxo group of thymine, and analogously the severe steric clash between the 2-amino group of diaminopurine and the 2-thio group of thiothymine

8.3.4 Double Duplex Invasion

Instead of aiming at stabilizing the PNA-DNA duplex to a level that enables strand displacement with only one duplex, one could also try to develop "pseudocomplementary" nucleobases that would allow for targeting of both DNA strands simultaneously. In this case the two PNAs would of course be sequence complementary, and therefore the nucleobase modifications should prevent the PNAs from recognizing each other without affecting their recognition of the complementary DNA. As also realized by Kutyavin et al. (1996) and tested in a DNA context, adenine to diaminopurine and simultaneous thymine to thiothymine substitution could fulfil these requirements (Fig. 13). Although the results using DNA were only moderately encouraging (Kutyavin et al. 1996), our experiments so far using PNA show that very efficient binding of a fully purine/pyrimidine mixed sequence PNA containing

60% diaminopurine-thiouracil base pairs to its complementary target in a double stranded DNA takes place by double duplex invasion (Lohse 1997; Lohse et al. 1998). Further experiments will explore the generality of this principle, reveal if it can be extended also to G-C base pairs, and show what are the biological effects on, e.g., the transcription and replication processes of such double duplex invasion complexes.

8.4 Concluding Remarks

Using either side-by-side minor groove binders or PNA it now seems practically possible to sequence selectively target virtually any sequence up to at least ten base pairs in double stranded DNA. The next challenge – apart from optimizing and refining these DNA binding ligands, and further characterizing their binding properties – will be to exploit such molecules for specific modulation of gene expression, and as molecular tools in gene technology. Several studies have already pointed out some of the directions, such as antigene therapy (Hanvey et al. 1992; Nielsen et al. 1994; Gottesfeld et al. 1997) and genome cutting and mapping procedures (Demidov et al. 1993, 1994; Veselkov et al. 1996a,b), but undoubtedly many more will be developed.

Acknowledgements. This work was supported by the Danish National Research Foundation.

References

Bentin T, Nielsen PE (1996) Enhanced peptide nucleic acid (PNA) binding to supercoiled DNA: possible implications for DNA "breathing" dynamics. Biochemistry 35:8863–8869

Cherny DY, Belotserkovskii, BP, Frank-Kamenetskii MD, Egholm M, Buchardt O, Berg RH, Nielsen PE (1993) DNA unwinding upon strand-displacement binding of a thymine-substituted polyamide to double stranded DNA. Proc Natl Acad Sci USA 90:1667–1670

Choo Y, Klug A (1995) Designing DNA-binding proteins on the surface of filamentous phage. Curr Opin Biotechnol 6:431–436

Choo Y, Klug A (1997) Physical basis of a protein-DNA recognition code. Curr Opin Struct Biol 7:117–125

Demidov V, Frank-Kamenetskii MD, Egholm M, Buchardt O, Nielsen PE
(1993) Sequence selective double strand DNA cleavage by peptide nucleic
acid (PNA) targeting using nuclease S1. Nucleic Acids Res 21:2103–2107

Demidov VV, Cherny DI, Kurakin AV, Yavnilovich MV, Malkov VA, Frank-
Kamenetskii MD, Sönnichsen SH, Nielsen PE (1994) Electron microscopy
mapping of oligopurine tracts in duplex DNA by peptide nucleic acid targeting. Nucleic Acids Res 22:5218–5222

Demidov VV, Yavnilovich MV, Belotserkovskii BP, Frank-Kamenetskii MD,
Nielsen PE (1995) Kinetics and mechanism of polyamide ("peptide") nucleic acid binding to duplex DNA. Proc Natl Acad Sci USA 92:2637–2641

Dervan PB(1992) Reagents for the site-specific cleavage of megabase DNA.
Nature 359:87–88

Egholm M, Buchardt O, Nielsen PE, Berg RH (1992) Peptide nucleic acids
(PNA). Oligonucleotide analogues with an achiral peptide backbone. J Am
Chem Soc 114:1895–1897

Egholm M, Christensen L, Dueholm KL, Buchardt O, Coull, J Nielsen PE
(1995) Efficient pH independent sequence specific DNA binding by
pseudoisocytosine-containing bis-PNA. Nucleic Acids Res 23:217–222

Eldrup AB, Dahl O, Nielsen PE (1997) A novel PNA monomer for recognition
of thymine in triple helix structures. J Am Chem Soc 119:11116–11117

Felsenfeld G, Davies DR, Rich A (1957) Formation of a three-stranded
polynucleotide molecule. J Am Chem Soc 79:2023–2024

Good L, Nielsen PE (1997) Progress in developing PNA as gene targeted
drugs. Antisense Nucleic Acid Drug Dev 7:431–437

Gottesfeld JM, Neely L, Trauger JW, Baird EE, Dervan PB (1997) Regulation
of gene expression by small molecules. Nature 387:202–205

Griffith MC, Risen LM, Greig MJ, Lesnik EA, Sprankle KG, Griffey RH,
Kiely JS, Freier SM (1995) Single and bis peptide nucleic acids as triplex-
ing agents: binding and stoichiometry. J Am Chem Soc 117: 831–832

Hanvey JC, Peffer NJ, Bisi JE, Thomson SA, Cadilla R, Josey JA, Ricca DJ,
Hassman, CF, Bonham MA, Au KG, Carter SG, Bruckenstein DA, Boyd
AL, Noble SA, Babiss LE (1992) Antisense and antigene properties of pep-
tide nucleic acids. Science 258:1481–1485

Hélène C (1993) Sequence-selective recognition and cleavage of double-helical DNA. Curr Opin Biotechnol 4:29–36

Hyrup B, Nielsen PE (1996) Peptide nucleic acids (PNA). Synthesis, properties and potential applications (review). Bioorg Med Chem 4:5–23

Kuhn H, Demidov V, Frank-Kamenetskii MD, Nielsen PE (1998) Kinetic sequence discrimination of cationic bis-PNAs upon targeting of double
stranded DNA. Nucleic Acids Res 26:582–587

Kutyavin IV, Rhinehart RL, Lukhtanov EA, Gorn VV, Meyer RB Jr, Gamper
HB Jr (1996) Oligonucleotides containing 2-aminoadenine and 2-

thiothymine act as selectively binding complementary agents. Biochemistry 35:11170–11176

LeDoan T, Perrouault L, Praseuth D, Habhoub N, Decout N, Thuong JL, Lhomme NT, Hélène C (1987) Sequence-specific recognition, photocrosslinking and cleavage of the DNA double helix by an oligo-[alpha]-thymidylate covalently linked to an azidoproflavine derivative. Nucleic Acids Res 15:7749–7760

Lohse J (1997) The principle of non-complementarity. PhD thesis, University of Copenhagen

Lohse J, Dahl O, Nielsen PE (1998) (in preparation)

Lown JW (1994) DNA recognition by lexitropsins, minor groove binding agents. J Mol Recogn 7:79–88

Moser HE, Dervan PB (1987) Sequence-specific cleavage of double helical DNA by triple helix formation. Science 238:645–650

Nelson HCM (1995) Structure and function of DNA-binding proteins. Curr Opin Genet Dev 5:180–189

Nielsen PE (1997) Design of sequence specific DNA binding ligands. Chem Eur J 3:505–508

Nielsen PE, Egholm M, Berg RH, Buchardt O (1991) Sequence-selective recognition of DNA by strand displacement with a thymine-substituted polyamide. Science 254:1497–1500

Nielsen PE, Egholm M, Berg RH, Buchardt O (1993) Peptide nucleic acids (PNA) DNA analogues with a polyamide backbone. In: Crook S, Lebleu B (eds) Antisense research and application. CRC Press, Boca Raton, pp 363–373

Nielsen PE, Egholm M, Buchardt O (1994) Evidence for (PNA)-2-DNA triplex structure upon binding of PNA to dsDNA by strand displacement. J Mol Recogn 7:165–170

Nielsen PE, Christensen L (1996) Strand displacement binding of a duplex-forming homopurine PNA to a homopyrimidine duplex DNA target. J Am Chem Soc 118:2287–88

Parks ME, Baird EE, Dervan PB (1996) Recognition of 5'-(A,T)GG(A,T)$_2$-3' sequences in the minor groove of DNA by hairpin polymides. J Am Chem Soc 118:6153–6159

Peffer NJ, Hanvey JC, Bisi JE, Thomson SA, Hassman CF, Noble SA, Babiss LE (1993) Strand-invasion of duplex DNA by peptide nucleic acid oligomers. Proc Natl Acad Sci USA 90:10648–10652

Pelton JG, Wemmer DE (1989) Structural characterization of a 2:1 distamycin Ad(CGCAAATTGGC) complex by two-dimensional NMR. Proc Natl Acad Sci USA 86:5723–5727

Praseuth D, Grigoriev M, Guieysse AL, Pritchard LL, Harel-Bellan A, Nielsen PE, Hélène C (1996) Peptide nucleic acids directed to the promoter of the α-chain of the interleukin-2 receptor. Biochim Biophys Acta 1309:226–238

Rhodes D, Schwabe JWR, Chapman L, Fairall L (1996) Towards an understanding of protein-DNA recognition. Philos Trans R Soc Lond B Biol Sci 351:501–509

Swalley SE, Baird EE, Dervan PB (1997) Discrimination of 5'-GGGG-3', %'-GCGC-3' and 5'-GGCC-3' sequences in the minor groove of DNA by eight-ring hairpin polyamides. J Am Chem Soc 119:6953–6961

Veselkov AG, Demidov VV, Frank-Kamenetskii MD, Nielsen PE (1996a) PNA as a rare genome-cutter. Nature 379:214

Veselkov AG, Demidov VV, Nielsen PE, Frank-Kamenetskii M (1996b) A new class of genome rare cutters. Nucleic Acids Res 24:2483–2487

Wade WS, Mrksich M, Dervan PB (1992) Design of peptides that bind in the minor groove of DNA at 5'(A,T)G(A,T)C(A,T)-3' sequences by a dimeric side-by-side motif. J Am Chem Soc 114:8783–8794

White S, Baird EE, Dervan PB (1997) On the pairing rules for recognition in the minor groove of DNA by pyrrole-imidazole polyamides. Chem Biol 4:569–578

Wittung P, Nielsen P, Nordén B (1997) Extended DNA-recognition repertoire of peptide nucleic acid (PNA): PNA-dsDNA triplex formed with cytosine-rich homo-pyrimidine PNA. Biochemistry 36:7973–7979

9 Solid Phase Libraries of Glycopeptide Templates in the Study of Complex Oligosaccharide-Receptor Interactions

M. Meldal, P.M. StHilaire, and K. Bock

9.1   Introduction to Protein Carbohydrate Interaction 169
9.2   Oligosaccharide Libraries 170
9.3   Glycopeptide Templates as Oligosaccharide Mimetics 171
9.4   Why Libraries of Glycopeptide Templates? 172
9.5   Influence of the Support on Solid Phase Libraries 172
9.6   Preparation and Analysis
        of Solid Phase Glycopeptide Template Libraries 173
9.7   Identification of Ligands
        from the Solid Phase Glycopeptide Library 175
9.8   Conclusions 180
References 180

9.1 Introduction to Protein Carbohydrate Interaction

Carbohydrate binding proteins are involved in myriad biological functions including communication and adhesion between cells, adhesion of bacteria or viruses, activation of the innate immune system, leukocyte rolling, hepatic clearing of aged serum proteins, and sorting of newly synthesized glycoproteins (Dwek 1996; Varki 1993). Based on their mode of binding they have been divided into three major groups. The E-, L-, and P-selectins and the galectins are of the calcium-dependent

C-type lectins, which bind their ligand mainly through coordination of two vicinal hydroxy groups of a single sugar moiety to a bound calcium ion in the carbohydrate recognition domain (CRD), and the surrounding sugars of the oligosaccharide ligand add to the binding specificity through relatively weak additional interactions. Due to the nature of this interaction the specificity of selectin binding is quite broad (Kretzschmar et al. 1997). However, the high in vivo activity observed with selectin binding has yet to be explained.

The calcium-independent receptors are involved in the clearing and sorting of glycoproteins. These include the mannose 6-phosphate receptors and the hepatic Gal/GalNAc specific receptors (Lee 1989; Tong et al. 1989). These receptors are truly multivalent and bind to the termini of bi-, tri-, or tetra-antennary N-linked oligosaccharides. The binding of the natural ligands is in the nanomolar to micromolar range and they are highly specific.

The third type are the collectins, a group of carbohydrate binding proteins remaining from the early development of life and mostly involved in the innate immune system (Miyamura et al. 1994). They are large surface or serum proteins composed of bundles of structural collagen stalks with trimer heads of calcium-dependent CRDs. The trimer heads of the mannose binding protein have been crystallized and the distance between the binding sites determined to be 53 Å (Weis and Drickamer 1994). The binding to simple mannose oligosaccharides is relatively weak and the activation of the complement cascade requires the protein to interact multivalently with large polymannans on foreign cell surfaces (Franzyk et al. 1996). Multivalent synthetic ligands for these proteins are therefore difficult to design and prepare. However, the CRDs may be targeted with monovalent high-affinity ligands (Wu et al. 1996).

9.2 Oligosaccharide Libraries

Carbohydrate libraries have been prepared both in solution and in solid phase with some success. The solution synthesis of libraries by simultaneous random α/β glycosylation of several hydroxyl groups of unprotected carbohydrates yielded mixtures which were screened for biological activity (Ding et al. 1995). However, the activities of these libraries were low in selectin binding assays and the deconvolution of the mixtures proved too difficult. The solid phase split and combine synthesis of ~1300 di- and trisaccharides as a "one bead one compound" library with labeling using the halo-aromatic tagging technique by Still et al. (Nestler et al. 1994) was more successful and yielded non-natural oligosaccharide ligands for *Burhinia purpurea* lectin (Liang et al. 1996). However, the affinity of the new ligands was similar to that of the original disaccharide.

9.3 Glycopeptide Templates as Oligosaccharide Mimetics

When glycopeptides were first introduced as mimics for oligosaccharides (Meldal et al. 1994b) it was envisaged that the saccharide part would provide the specificity of the binding by directing the ligand to the oligosaccharide binding site while the peptide would function as a scaffold for optimal orientation of the glycan. Peptide ligands are known to generally bind with high affinity to their receptors and it is expected that the glycopeptide would furthermore interact favorably with the receptor through the peptide scaffold, leading to increased binding affinity (Christensen et al. 1994). The principle was consolidated through binding studies with a library of phosphorylated glycopeptides and the divalent mannose 6-phosphate receptor. Ligands binding 20-fold stronger than the natural phosphorylated pentamannose ligand were found. It was demonstrated that a minimum of disaccharides were required. Attempts to increase the affinity through cyclization were unsuccessful (Franzyk et al. 1997).

Similar results were later obtained with glycopeptides and binding to selectins. A glycopeptide mimic of the sialyl-Le$_x$ tetrasaccharide containing fucose on a peptide scaffold mimicking the carboxylate of the neuraminic acid and one hydroxyl group of the central galactose had a 60-fold increased binding affinity for E-selectin (Lin et al. 1996b). There was no increase in affinity observed when the ligands were immobilized in a polymeric multivalent arrangement in liposomes (Lin et al. 1996a).

A high-affinity divalent ligand to the adhesin of *Streptococcus suis* has been prepared containing Gal-α-1–4-Gal-α-linked via peptide bonds to an aromatic nucleus. The interaction was found to be truly divalent since the preparation of putative tetravalent ligands showed no significant increase in binding (Hansen et al. 1997).

The crystal structure of the complex between the cholera toxin and its pentasaccharide ligand showed that the terminal Gal residue was bound tightly to the receptor. A library of 89 scaffolds containing galactose was prepared in parallel by base-catalyzed Michael addition of β-D-$(C_{12}H_{25}CO)_4$Gal-SAc to ten different unsaturated ketones followed by reductive amination and N-acylation with a variety of acids. The galactose provided the specificity for the cholera toxin while the affinity was obtained through interaction with the aglyconic scaffold. In this way, 40 nM inhibitors of the toxin binding were obtained (O. Hindsgaul, personal communication).

9.4 Why Libraries of Glycopeptide Templates?

The affinity of glycopeptides to carbohydrate binding proteins may be utilized to identify high-affinity ligands in a library format. The ease with which glycopeptides are synthesized using preactivated amino acids and glycosylated amino acid building blocks can by careful assembly of a library ensure the generation of a single compound in each bead. For the preparation of glycosyl amino acid building blocks, the glycosylation of Fmoc-amino acid-OPfp esters has proved a general and versatile method useful for the preparation of complex compounds for direct incorporation into the glycopeptide libraries (Meldal and Bock 1994).

9.5 Influence of the Support on Solid Phase Libraries

Most solid supports for peptide synthesis including Tentagels and Polyhipe are based on polystyrene materials which absorb light, thus interfering with, e.g., fluorescence assays, and the hydrophobic core may result in non-specific protein binding. The polystyrene-based gels furthermore exclude biomolecules from their interior due to the hydrophobic core polymer (Vagner et al. 1996). Recently, novel types of gel supports based on polyethylene glycol (PEGA and POEPOP) were introduced (Auzanneau et al. 1995; Meldal 1992; Renil and Meldal

1996). These PEG polymers are cross-linked with long chain PEG macromonomers, and PEG chains present amino or hydroxyl functional groups. They are obtained by radical initiated inverse suspension polymerization of partially acryloylated PEG and by anion catalyzed bulk polymerization of PEG, derivatized with epichlorohydrine, respectively. The inert character of the polymers allows the application of organic reactions. Because of the excellent swelling behavior in aqueous buffers, PEGA resins have been used in bioassays for enzymes and in protein binding studies with no non-specific binding of the proteins (Meldal et al. 1994a,c; Meldal and Srendsen 1995; Spetzler et al. 1997). The PEGA polymer beads were therefore employed in the present work for the generation of glycopeptide libraries.

9.6 Preparation and Analysis of Solid Phase Glycopeptide Template Libraries

Quantitative peptide assembly ensures the necessary purity for the reproducible identification of active compounds in the library by analytical techniques. Since the deconvolution of the library is possible by structural analysis, the labeling by chemical tagging can in principle be omitted. However, the structural analysis of picomolar amounts of compound present on a single bead is still by no means routine. Peptides on beads can be conveniently analyzed by solid phase ladder sequencing (Chait et al. 1993). However, the glycopeptides were not stable under the conditions of Edman degradation cycles and the glycans were partly lost giving rise to multiple fragments which were difficult to identify.

Alternatively, synthetic history can be captured in the beads by capping in each synthetic step (Youngquist et al. 1995). Thus, a series of related fragments rather than a single compound are generated in the bead. The difficulty arises when this technique is applied in a glycopeptide library in which the amino nucleophiles have a high variety of reactivity. Both post- and precapping techniques are doomed to fail. Even the use of a simple in situ capping agent in mixture with the activated amino acids is problematic resulting in anywhere from 0% to 100% capping. The method that was eventually developed is based on the use of structurally closely related capping agents and building blocks in mixtures of 90% Fmoc-amino acid-OPfp and 10% of the same

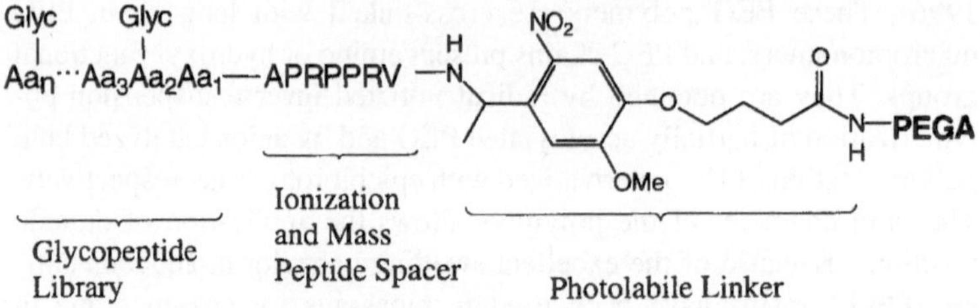

Fig. 1. The structure of the glycopeptide library was designed for easy identification of active glycopeptides by mass spectrometry

Boc-amino acid-OPfp esters (or the free acids activated by TBTU) for all the couplings of natural amino acids.

The couplings of glycosylated Fmoc-amino acid-OPfp esters were encoded by capping with selected carboxylic acids-OPfp esters with masses different from those of the regular amino acids. First the reactivity of the carboxylic acids was determined by reaction of a mixture of Fmoc-Thr(Ac$_4$Man-α)-OPfp and each of the carboxylic acid OPfp esters with a peptide resin. The reactivity ranged from seven times less to several times higher than that of the glycosylated amino acid-OPfp ester and the ratio for coupling in the libraries was adjusted accordingly.

The library was linked to the polymer via the photolabile linker presented in Fig. 1, which is cleaved at 351 nm by the nitrogen laser used in matrix-assisted laser-desorption ionization time-of-flight (MALDI-TOF) mass spectrometers. The linker facilitated the immediate analysis of compounds released from the resin beads after application of these to the laser target in the presence of α-cyano-4-hydroxycinnamic acid (CHC) matrix.

The synthesis of the libraries was performed in an MCPS Teflon library generator with 20 columns and a mixing chamber above the columns (Meldal 1994). During mixing, the columns and half the volume of the chamber were filled with solvent, a circular lid sealed with an O-ring was fitted and the resin thoroughly agitated for 15 min. Washing solvents and piperidine solution for deprotection were added from dispenser bottles connected to 20 line dispenser heads mounted on an aluminum frame. During couplings, the solutions of Pfp esters were pipetted into the MCPS reaction columns and the reactor was agitated for 3 h allowing three couplings per working day. Piperidine (20%) in DMF was used for deprotection of the α-amino group in each reaction cycle and at the end of the assembly the library was deprotected during a 2-h treatment with a mixture of TFA and scavengers. The sugar O-acetates were removed with hydrazine hydrate in methanol during a 2-h treatment.

The purity of the library was assessed by MALDI-TOF mass spectrometry by collection and analysis of a few beads. Most of the beads collected afforded spectra as ladders, which could easily be deciphered using mass difference assignment software from Bioanalysis.

9.7 Identification of Ligands
from the Solid Phase Glycopeptide Library

The library and resin containing no peptide (negative control) were incubated with the fluorescence labeled lectin from *Lathyrus odoratus*. The control beads were not fluorescent whereas the glycopeptide library provided a small range of more or less fluorescent beads against a background of non-fluorescent beads. The most fluorescent beads were collected, crushed on a target, mixed with CHC matrix and subjected to MALDI-TOF mass spectrometry. One-third of the beads contained no glycan and were pure peptides. Two-thirds of the beads contained either mannose, *N*-acetylglucosamine or both (St. Hilaire et al. 1998).

The active sequences were synthesized and binding of the lectin to resin-bound glycopeptides revealed the two most active peptides identified to be H-T(Man)FFFVNKV-NH$_2$ and H-T(Man)LFKGFHV-NH$_2$. They both contained a mannose α-linked to threonine at the N-terminal of the peptide. They also contained aromatic F and H residues in agreement with previous peptide mimics as ligands for oligosaccharide binding proteins. Glycopeptides with Man or GlcNAc linked to central amino acid residues bound with less affinity. Mannose and mannan at high concentration were able to competitively inhibit the binding of the receptor to T(Man)FFFVNKV linked to the PEGA beads.

A similar approach was employed to identify oligosaccharide receptors from porcine liver. The liver acetone powder was treated as described by Kornfeldt et al. and the protein from the pellet was resuspended and labeled with fluorescein isothiocyanate. A ladder-encoded

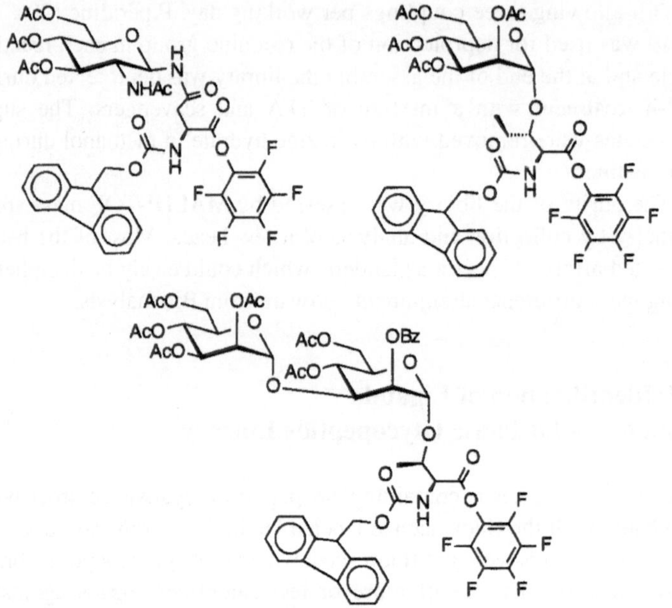

Fig. 2. The building blocks employed in the construction of glycopeptide libraries with labeling using carboxylic acids: 2-phenylpropanoic acid, 2-naphthoic acid and dodecanoic acid

glycopeptide library was synthesized on PEGA resin using a photolabile linker, a -VPRPPRV- peptide mass spacer and either the three building blocks shown in Fig. 2 with encoding and the procedure described above or five building blocks including also Fmoc-S(Bz$_4$Glc-β)-OPfp and Fmoc-Y(Ac$_4$Gal-β)-OPfp and no glycosyl amino acid encoding. The libraries were incubated for 1 h with the extract mixture of porcine liver proteins which had been labeled with fluorescein. Approximately 1 out of 500 of the beads was highly fluorescent and an assembly of such beads was collected and subjected to structural analysis by MALDI-TOF mass analysis. This yielded a range of glycopeptides in addition to some non-glycosylated structures. Examples of the high-resolution spectra obtained with a Bruker Reflex III instrument are presented in Fig. 3. The presence of non-glycosylated peptides was not surprising since peptides are known to bind oligosaccharide receptors

Table 1. Structures found to bind the Man/GlcNAc lectin isolated from porcine liver acetone powder. Some of the synthesized glycopeptides were found to bind the Man/GlcNAc receptor in a competitive manner in a binding inhibition assay as will be described in detail elsewhere. Three types of glycopeptides were identified, those containing both Man and GlcNAc and those containing either Man or GlcNAc. No recognition of Gal containing peptides was observed

A	S	Ng	Y	W	S	Tm6		Type 1
P	E	Ng	L	F	F	Tm	♣	
N/D	Tm	P	E	Ng	Y	K		
Ng	F	T	K	F	Tm3	E	♣	
A	A	D	V	A	Ng	E		Type 2
Ng	G	F	Y	V	D	P	♣	
Ng	H	P	L	F	L	F		
E	Ng	W	T	L	G	F		
A	Ng	L	P	T	K	W		
A	Ng	Y	P	V	T	L		
V	P	Ng	F	F	W	F		
E	V	F	Ng	F	S	E	♣	
F	E	V	T	F	E	Tm3	♣	Type 3
–	P	V	Tm	T	Tm	Y	♣	

Ng=Asn(β-*N*-GlcNAc-); Tm=Thr(α-*O*-Man-); Tm6/3=Thr(α-Man-1-3/6-α-*O*-Man-); ♣, synthesized.

and furthermore the crude preparation from porcine liver could contain other proteins with affinity for peptides. Although galactose was also included in one of the libraries, only glycopeptides containing GlcNAc and man showed fluorescence (Table 1) in agreement with the recent identification of a man/GlcNAc specific hepatic lectin in rat liver. Some of the identified glycopeptides were resynthesized and their capacity to inhibit receptor binding to ligands on the beads was compared in inhibition studies.

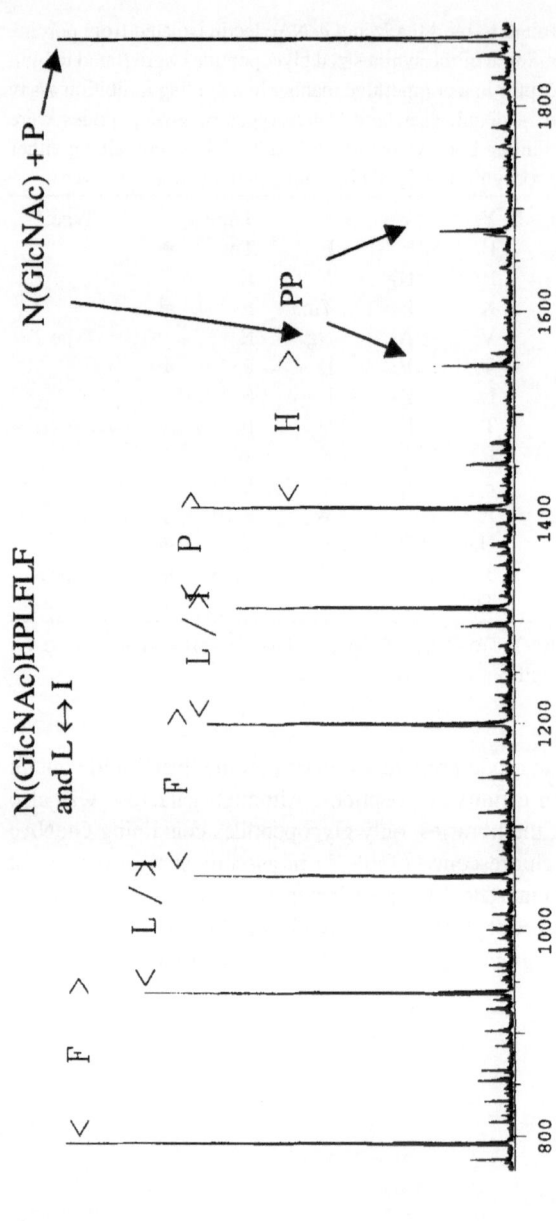

Fig. 3. Legend see p. 179

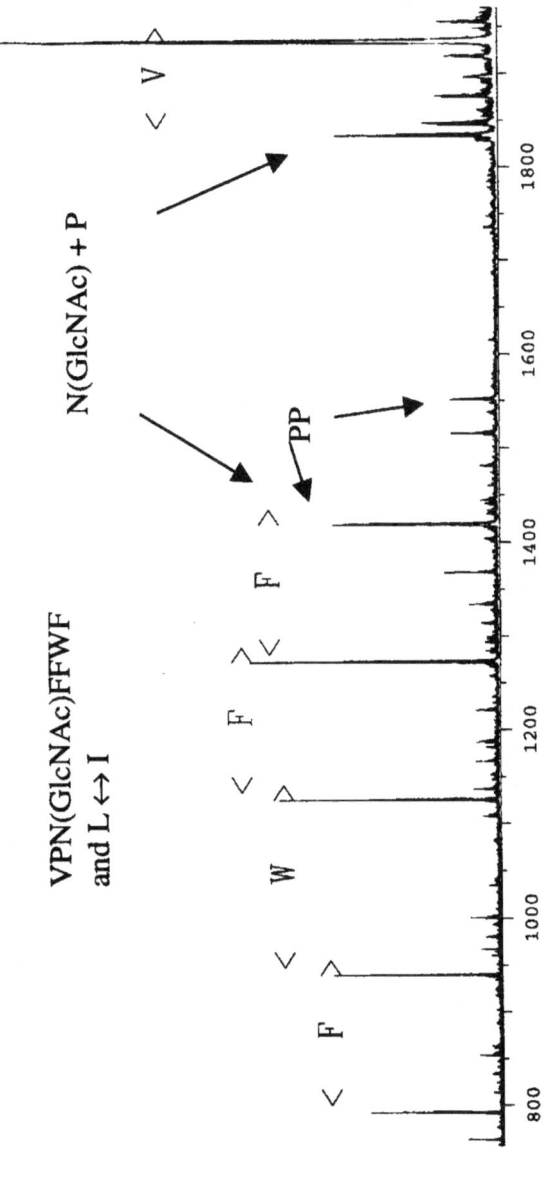

Fig. 3. MALDI TOF-mass spectra (ReflexIII, Bruker) from direct photolytic release of active ligands from the PEGA beads. The ligands were prepared by ladder synthesis using in situ capping with a mixture of the Fmoc- and Boc-amino acids. Glycans were encoded by carboxylic acid labels with the additional molecular mass: 132.2 (2-phenylpropanoic acid), 154.2 (2-naphthoic acid) and 182.3 (dodecanoic acid). High accuracy and unambiguous assignment could be obtained. The two active sequences shown contained GlcNAc and hydrophobic amino acids

9.8 Conclusions

A novel method has been presented for the identification of glycopeptide ligands for oligosaccharide binding proteins from a solid phase glycopeptide library. The method utilizes in situ capping in each synthesis step with a mixture of Boc and Fmoc amino acids while glycosylated amino acids are separately encoded by capping with carboxylic acids. The analysis of structures is performed by direct photolytic release from PEGA support in the MALDI-TOF mass spectrometer. Application of this technique yields high-affinity ligands for carbohydrate binding proteins and can be used to identify new carbohydrate binding proteins.

Acknowledgement. The work has been supported by the Mitzutani foundation and the EU-Science program (grant SCI-CT 92-0765).

References

Auzanneau F-I, Meldal M, Bock K (1995) Synthesis, characterization and biocompatibility of PEGA resins. J Pept Sci 1:31–44

Chait BT, Wang R, Beavis RC, Kent SBH (1993) Protein ladder sequencing. Science 262:89–92

Christensen MK, Meldal M, Bock K, Cordes H, Mouritsen S, Elsner H (1994) Synthesis of glycosylated peptide templates containing 6'-O- phosphorylated mannose disaccharides and their binding to the cation-independent mannose-6-phosphate receptor. J Chem Soc Perkin Trans 1:1299–1310

Ding Y, Kanie O, Labbe J, Palcic MM, Ernst B, Hindsgaul O (1995) Synthesis and biological activity of oligosaccharide libraries. In: Alavi A, Axford JS (eds) Glycoimmunology. Plenum, New York, p 261

Dwek RA (1996) Glycobiology: toward understanding the function of sugars. Chem Rev 96:683–720

Franzyk H, Meldal M, Paulsen H, Thiel S, Jensenius JC, Bock K (1996) Glycopeptide mimics of mammalian Man(9)GlcNAc(2) – ligand-binding to mannan-binding proteins (MBPs). Bioorg Med Chem 4:1881–1899

Franzyk H, Christensen MK, Jørgensen RM, Meldal M, Cordes H, Mouritsen S, Bock K (1997) Constrained glycopeptide ligands for MPRs – limitations of unprotected phosphorylated building-blocks. Bioorg Med Chem 5:21–40

Hansen HC, Haataja S, Finne J, Magnusson G (1997) Di-, tri, and tetravalent dendritic galabiosides that inhibit hemagglutination by Streptococcus suis at nanomolar concentration. J Am Chem Soc 119:6974–6979

Kretzschmar G, Toepfer A, Hüls C, Krause M (1997) Pitfalls in the synthesis and biological evaluation of sialyl-Lewis[x] mimetics as potential selectin antagonists. Tetrahedron 53:2485–2494

Lee YC (1989) Binding modes of mammalian hepatic Gal/GalNAc receptors. In: Bock G, Harnett S (eds) Carbohydrate recognition in cellular function. Wiley, Chichester, p 80

Liang R, Yan L, Loebach J, Ge M, Uozume Y, Sekanina K, Horan N, Gildersleeve J, Thompson C, Smith A, Biswas K, Still WC, Kahne D (1996) Parallel synthesis and screening of a solid phase carbohydrate library. Science 274:1520–1522

Lin CC, Kimura T, Wu SH, Weitzschmidt G, Wong CH (1996a) Liposome-like fucopeptides as sialyl-Lewis-X mimetics. Bioorg Med Chem Lett 6:2755–2760

Lin C-C, Shimazaki M, Heck M-P, Aoki S, Wang R, Kimura T, Ritzen H, Takayama S, Wu S-H, Weitz-Schmidt G, Wong C-H (1996b) Synthesis of sialyl Lewis X mimetics and related structures using the glycosyl phosphite methodology and evaluation of E-selectin inhibition. J Am Chem Soc 118:6826–6840

Meldal M (1992) PEGA: a flow stable polyethylene glycol dimethyl acrylamide copolymer for solid phase synthesis. Tetrahedron Lett 33:3077–3080

Meldal M (1994) Multiple column synthesis of quenched solid-phase bound fluorogenic substrates for characterization of endoprotease specificity. Methods (a companion to Methods Enzymol) 6:417–424

Meldal M, Bock K (1994) A general-approach to the synthesis of O-linked and N-linked glycopeptides. Glycoconjugate J 11:59–63

Meldal M, Svendsen I (1995) Direct visualization of enzyme inhibitors using a portion mixing inhibitor library containing a quenched fluorogenic peptide substrate. 1: inhibitors for subtilisin Carlsberg. J Chem Soc Perkin Trans 1 1591–1596

Meldal M, Auzanneau F-I, Hindsgaul O, Palcic MM (1994a) A PEGA resin for use in solid phase chemical/enzymatic synthesis of glycopeptides. J Chem Soc Chem Commun 1849–1850

Meldal M, Christiansen-Brams I, Christensen MK, Mouritsen S, Bock K (1994b) Synthesis and biological application of glycosylated peptide templates. In: Bock K, Clausen H (eds) Complex carbohydrates in drug research: structural and functional aspects. Copenhagen, Munksgaard, p 153

Meldal M, Svendsen I, Breddam K, Auzanneau FI (1994c) Portion-mixing peptide libraries of quenched fluorogenic substrates for complete subsite mapping of endoprotease specificity. Proc Natl Acad Sci USA 91:3314–3318

Miyamura K, Reid KMB, Holmskov U (1994) The collectins – mammalian lectins containing collagen-like regions. Trends Glycosci Glycotechnol 6:286–309

Nestler HP, Bartlett PA, Still WC (1994) A general method for molecular tagging of encoded combinatorial chemistry libraries. J Org Chem 59:4723–4724

Renil M, Meldal M (1996) POEPOP and POEPS: inert polyethylene glycol crosslinked polymeric supports for solid phase synthesis. Tetrahedron Lett 37:6185–6188

Spetzler JC, Westphal V, Winther JR, Meldal M (1998) Preparation of fluorescence quenched libraries containing inter-chain disulfide bonds for studies of protein disulfide isomerases. J Pept Sci 4:128–137

St Hilaire PM, Lowary TL, Meldal M, Bock K (1998) Oligosaccharide mimetics obtained by novel, rapid screening of carboxylic acid encoded glycopeptide libraries. J Am Chem Soc (submitted)

Tong PY, Gregory W, Kornfeld S (1989) Ligand interactions of the cation-independent mannose 6-phosphate receptor. J Biol Chem 264:7962–7969

Vagner J, Barany G, Lam KS, Vágner J, Barany G, Lam KS, Krechnak V, Sepetov NF, Ostrem JA, Strop P, Lebl M (1996) Enzyme-mediated spatial segregation on individual polymeric support beads: application to generation and screening of encoded combinatorial libraries. Proc Natl Acad Sci USA 93:8194–8199

Varki A (1993) Biological roles of oligosaccharides: all of the theories are correct. Glycobiology 3:97–130

Weis WI, Drickamer K (1994) Trimeric structure of a C-type mannose-binding protein. Structure 2:1227–1240

Wu S-H, Shimazaki M, Lin C-C, Qiao L, Moree WJ, Weitz-Schmidt G, Wong C-H (1996) Synthesis of fucopeptides as sialyl Lewisx mimetics. Angew Chem Int Ed Engl 35:88–90

Youngquist RS, Fuentes GR, Lacey MP, Keough T (1995) Generation and screening of combinatorial peptide libraries designed for rapid sequencing by mass spectrometry. J Am Chem Soc 117:3900–3906

10 The Molecular Recognition of Saccharides and Glycoprotein-Inspired Materials

L.L. Kiessling

10.1   Protein–Carbohydrate Interactions Mediate
      Important Biological Recognition Events 184
10.2   Synthesis of Multivalent Displays of Saccharides Using ROMP   186
10.3   Cooperative Intermolecular Interactions: The Chelate Effect 189
10.4   Translational and Conformational Entropy 191
10.5   Statistical Effects of High Binding Site Density 192
10.6   Extended and Secondary Binding Sites 193
10.7   Steric Effects 194
10.8   Synthetic Scaffolds for Multivalent Presentation 195
10.9   Structure/Function Studies of Neoglycopolymer Inhibitors
      of Concanavalin A-Mediated Cell Agglutination 196
10.10   Multivalent Ligand Specificity:
      Structure of the Recognition Elements 196
10.11   Multivalent Synthetic Ligands Differing in Average Lengths 198
10.12   Recognition of Multivalent Ligands
      by Different Saccharide-Binding Proteins 203
10.13   Conclusions 206
References 207

10.1 Protein–Carbohydrate Interactions Mediate Important Biological Recognition Events

Studies of biological molecular recognition events have primarily focused on high affinity recognition processes. Our research program
focuses instead on a class of apparent weak interactions that play significant roles in biology, those that occur when proteins bind to extracellular
carbohydrates. Such protein–carbohydrate interactions are essential participants in many physiological cell–cell recognition processes, including fertilization, bacterial and viral pathogenesis, and the inflammatory
response (Dwek 1996; Feizi 1993; Gabius 1997; Lis and Sharon 1998;
Varki 1993). Consistent with their roles in fundamental biological processes, saccharides display many functional groups on diverse scaffolds.
Indeed, oligosaccharides can display a diversity of structure that far
exceeds that of proteins and nucleic acids (Laine 1994). Even with this
potential for conveying structural information, monovalent protein-carbohydrate interactions often occur not only with low affinity (i.e., $K_a \approx$
$10^{3–4}$ M^{-1}) (Elgavish and Shaanan 1997; Glaudemans 1991; Lee and
Lee 1995; Lis and Sharon 1998; Quiocho et al. 1989; Weis and
Drickamer 1996) but also with broad specificity. We have begun to
address key questions in this area: How do low affinity protein – saccharide interactions mediate specific molecular recognition events? What
advantages do these interactions confer in a biological setting?

In physiological settings, saccharide epitopes are not usually encoun-
tered in isolation but rather in multivalent arrays (Fig. 1). For example,
membrane-bound glycoproteins can serve as scaffolds that present multiple copies of oligosaccharide determinants. Alternatively, saccharide
recognition elements can be displayed as clustered glycolipids. Moreover, many saccharide-binding proteins form dimeric, trimeric or tetra-
meric or oligomeric quaternary structures (Drickamer 1995; Rini 1995;
Weis and Drickamer 1996). These observations have led to the specula-
tion and, in some cases demonstration, that the presentation of multiple
copies of saccharide recognition elements plays a critical recognition
role (Kiessling and Pohl 1996; Lee and Lee 1995; Roy 1996). In these
milieux, the functional affinity and selectivity of the interaction can be
augmented (Fig. 1). Consequently, to understand how saccharide structures convey information, platforms that exhibit recognition elements in
multivalent arrays are needed (Kiessling and Pohl 1996; Liang et al.

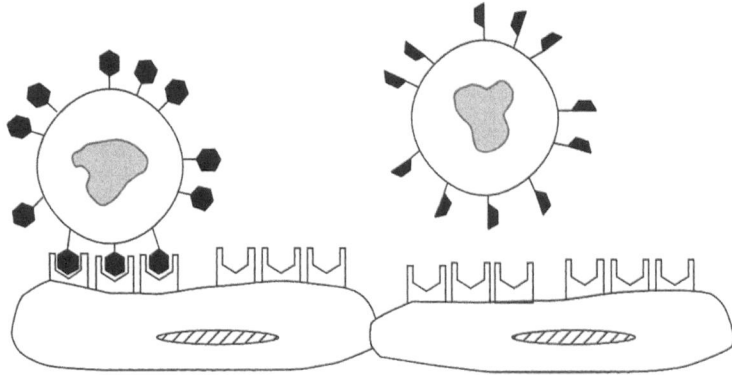

Fig. 1. Carbohydrates presented as glycolipids or glycoproteins (represented by *black shapes*) are displayed in clusters on the cell surface. Multipoint binding between carbohydrate-binding proteins on the cell surface and saccharide displays is one mechanism that can be used to augment the affinity and specificity of a cell–cell interaction

1997; Roy et al. 1996). With this model in mind, we have developed strategies to synthesize materials that mimic the putative multivalent ligands. The information gained from such studies can be used to the design molecules that can illuminate, inhibit, or exploit multidentate binding.

To emulate natural multivalent carbohydrate displays, natural and non-natural macromolecular scaffolds for saccharide residue presentation have been generated, including neoglycoproteins, neoglycolipids, amino acid homopolymers, liposomes, polyacrylamides, dendrimers, and low molecular weight (e.g., ≤1000) displays (Kiessling and Pohl 1996; Lee and Lee 1995; Roy et al. 1996). Small scaffolds are attractive as their defined structure could facilitate structure/function studies and they may serve as drug leads (Glick and Knowles 1991; Hansen et al. 1997; Lee and Lee 1987). Unfortunately, many of the small, defined displays of multivalent ligands are only slightly more active than the monovalent ligand, and they are difficult to synthesize. In contrast, neoglycolipids (Feizi et al. 1994) and neoglycoproteins (Wong 1995) often afford large increases in functional affinities relative to monovalent ligands. Still, issues such as the spacing between recognition elements cannot be addressed with these materials, because the locations and geometries of the saccharide residues are ill defined (Boullanger 1997). Polymeric displays also often offer large (up to 10^7-fold) enhancements in functional affinities, and they are readily synthesized (Bovin and Gabius 1995; Kiessling and Pohl 1996; Mammen et al. 1995; Roy et al. 1996; Schnarr et al. 1979; Spaltenstein and Whitesides 1991). In addition, polymerization strategies also provide the means to engineer useful features into the structure of a multivalent ligand.

Not all polyvalent displays, however, are potent inhibitors. The mode of saccharide epitope display may be critical for tight binding to the target protein, but limited information has been gathered on these systems. As with neoglycoproteins and neoglycolipids, a barrier to elucidating the underlying molecular mechanisms that contribute to increases in avidity for polyvalent ligands is the difficulty of generating molecules in which saccharide presentation is systematically controlled.

Our goal was to develop a method to generate multivalent materials that have the favorable binding features of the polyvalent displays, and that could be used to elucidate the mechanisms by which multivalent ligands function. Assembly methods that can be used to engineer the scaffold structure upon which the multivalent ligands are displayed are of significant value, because they provide the means to generate mimics of cell surface glycoproteins and glycolipid displays. We hypothesized that the ring-opening metathesis polymerization (ROMP) could be used for this purpose.

10.2 Synthesis of Multivalent Displays of Saccharides Using ROMP

The demonstration that ROMP could be used to create molecules for exploration and/or modulation of biological recognition processes was lacking. Our immediate goal, therefore, was to rapidly prepare carbohydrate-substituted materials to determine whether they could serve as ligands for carbohydrate-binding proteins. To this end, we synthesized glucose-substituted monomer **1** and polymerized it using ruthenium trichloride in aqueous solutions to afford multivalent materials in very good yields (ca. 80%–85%) (Fig. 2) (Mortell et al. 1994). Gel filtration analysis of the resulting polymers suggested they had relative molecular

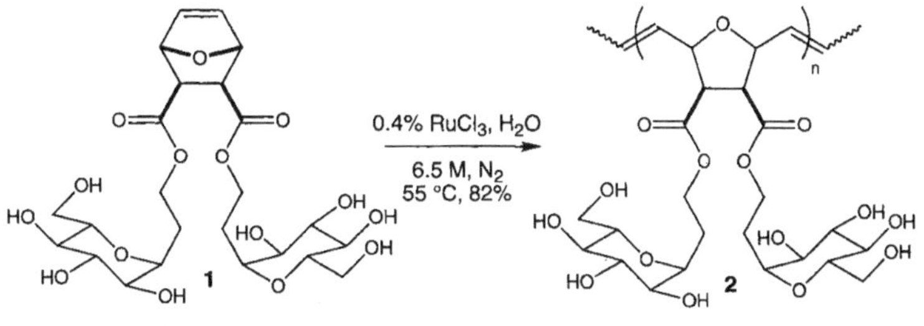

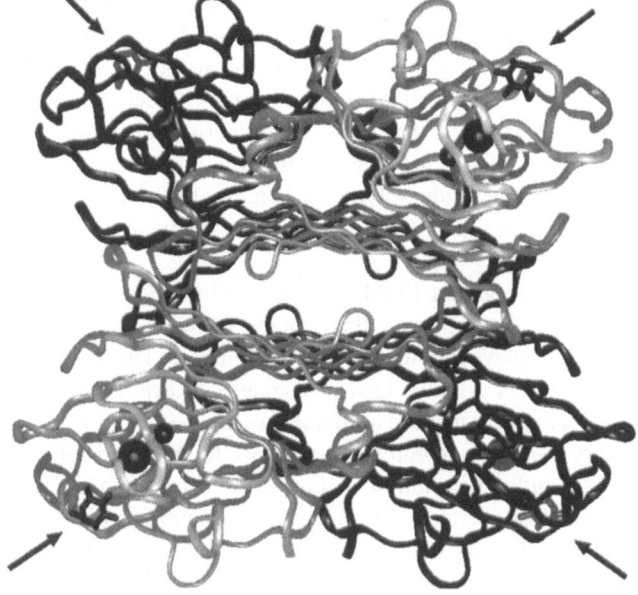

Fig. 2. The ring-opening metathesis polymerization provides access to carbohydrate-substituted materials. A *C*-glycoside polymer can be generated from the bicyclic 7-oxanorbornene template using ruthenium trichloride in aqueous solution

Binding Sites Separated by 65 Å

Fig. 3. The structure of the concanavalin A tetramer bound to mannose as determined by x-ray crystal structure analysis. The *spheres* represent the cationic metal ions that play important structural roles. The *arrows* indicate the saccharide binding sites

masses (M_r) in the range of 10^6. These studies, and those of others (Fraser and Grubbs 1995; Nomura and Schrock 1996), demonstrated that ROMP could be used to generate saccharide-substituted materials.

To determine whether these materials could act as bioactive ligands, they were tested in a cell agglutination assay using the saccharide-binding protein concanavalin A. Concanavalin A is a plant lectin that binds both glucose and mannose residues (Bittiger and Schnebli 1976). The structure of concanavalin A complexed with mannose derivatives has been determined by x-ray crystallographic studies (Fig. 3) (Derewenda et al. 1989; Loris et al. 1996). Because concanavalin A exists as a homotetramer at neutral pH, it can bind glycoproteins on opposing red blood cell surfaces thereby acting like a molecular glue that holds cells together. The interaction of monovalent and multivalent ligands with concanavalin A, therefore, can be assessed in a hemagglutination assay (Osawa and Matsumoto 1972), in which the abilities of ligands to inhibit concanavalin A-facilitated agglutination of red blood cells is evaluated. To determine the efficacies of the various compounds, we compared the abilities of monovalent ligands and the multidentate polymers to inhibit concanavalin A-mediated cell agglutination on a saccharide residue basis. Our studies revealed that the first saccharide-substituted molecules generated by ROMP are potent inhibitors of cell agglutination. Specifically, the C-glucoside polymer 2 is a 2000-fold more effective inhibitor of concanavalin A than is monovalent α-methyl glucopyranoside (Mortell et al. 1994).

These studies reveal that ROMP can be used to generate materials that present saccharide epitopes that can be recognized by proteins. Moreover, such displays can serve as potent inhibitors. The mechanisms by which multivalent displays lead to increases in avidity are largely unknown. With this new method to rapidly assemble such arrays, we set out to identify molecular features that give rise to enhanced functional affinities. Such studies will illuminate important features of the molecular recognition of saccharides and assist in the formulation of strategies for the design of inhibitors of carbohydrate-binding proteins. In the next section, several mechanisms that can operate in multivalent binding events are discussed as a backdrop for interpreting our studies of saccharide binding and function.

10.3 Cooperative Intermolecular Interactions:
The Chelate Effect

The interaction of oligomeric proteins with multidentate saccharide derivatives can lead to high avidity interactions when the orientation of the carbohydrate recognition elements corresponds to that of the binding sites within the lectin (carbohydrate-binding protein) subunits. For example, the free energy of interaction ($\Delta G_{observed}$) for a dimeric receptor (R_1–R_2) binding to a multivalent ligand (L_1–L_2–L_3–L_4) will be related to the sum of the intrinsic free energies of binding for each individual receptor-ligand pair (Crothers and Metzger 1972; Jencks 1981; Page and Jencks 1971). For the overall binding event depicted in Fig. 4A, which involves identical receptor subsites binding to identical recognition elements, the sum of the intrinsic free energies of binding is often used as an approximation of the observed binding free energy. In the specific example depicted in Fig. 4A, the overall free energy of binding would be estimated by Eq. 1.

$$\Delta G_{observed} = \Delta G(R_1\text{-}L_2) + \Delta G(R_2\text{-}L_3) \tag{1}$$

$\Delta G_{observed}$ can also be approximated by $2\Delta G_{R\text{-}L}$, where $\Delta G_{R\text{-}L}$ is the free energy of binding for a single receptor-ligand pair. A generalized form of this approximation for a system in which n identical receptor–ligand pairs can form is depicted as Eq. 2.

$$\Delta G_{observed} = \Delta G(R_1\text{-}L_1) + \Delta G(R_2\text{-}L_2) +... + \Delta G(R_n\text{-}L_n)$$
$$\text{or } n\Delta G(R\text{-}L) \tag{2}$$

The model predicts that the formation of additional contacts should be facile as only one translational entropy penalty is paid upon formation of the first receptor–ligand interaction (Jencks 1981; Page and Jencks 1971). Still, this formulation does not address the conformational entropy penalty, which may be significant in some systems (vide infra).

The idea that high functional affinities can be achieved through the formation of multiple receptor–ligand contacts is consistent with the chelate effect (Jencks 1981; Page and Jencks 1971) and with models for antibody–antigen interactions (Bell 1974; Crothers and Metzger 1972). High functional affinities for these multivalent receptor ligand interac-

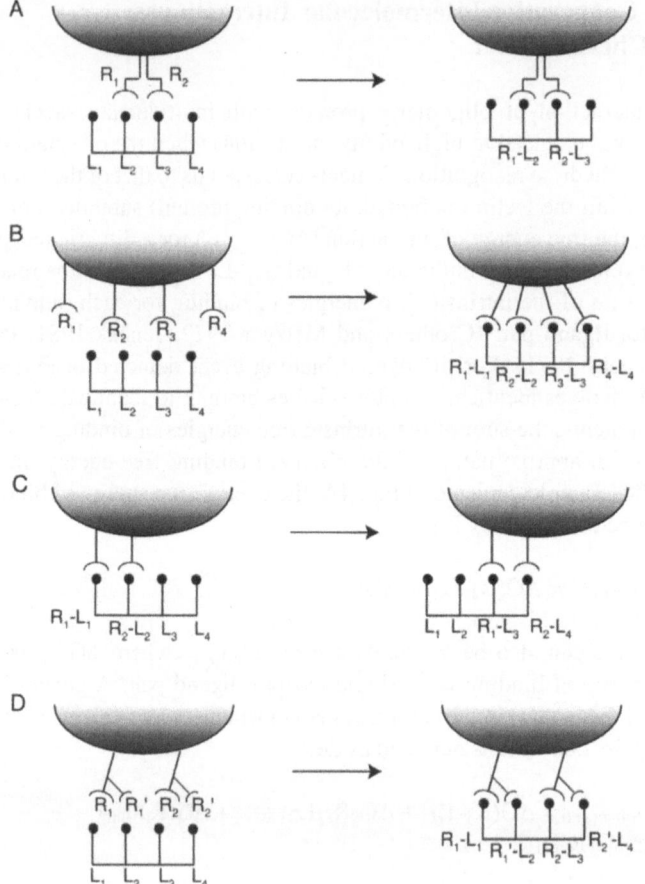

Fig. 4A–D. Possible molecular mechanisms for increases in functional affinities for multivalent ligands. **A** Interaction of a multivalent ligand with a multivalent receptor (chelate effect). **B** Clustering of cell surface receptors by a multivalent ligand. **C** Increased local concentration of recognition epitopes leads to slow off-rates (statistical effect). **D** Subsite binding of saccharide residues that involves occupation of secondary binding sites on a single receptor

tions are predicted, even for binding events that involve formation of only a few individual receptor–ligand complexes. Studies of physiological multivalent protein–saccharide interactions suggest that the chelate effect is important (Kiessling and Pohl 1996; Lee and Lee 1995; Roy 1996). Additionally, a non-natural trimer of the glycopeptide antibiotic vancomycin binds a trimer of its dipeptide ligand D-Ala-D-Ala with a functional affinity similar to that of the avidin–biotin interaction (Rao et al. 1998). In natural settings, the strength of binding mounted through multivalent interactions may also be affected by several other parameters outlined below.

10.4 Translational and Conformational Entropy

Translational and conformational entropy terms play important roles in determining the functional affinity of a particular interaction. For multidentate ligands, the cost for loss of translational entropy is paid with formation of the first individual interaction, and the probability that subsequent complexes will form from additional available binding subsites is increased because of the high effective concentration of recognition elements (Jencks 1981; Page and Jencks 1971). This favorable translational entropy contribution for multivalent binding is opposed by a conformational entropy penalty. The latter term is affected by features of the multivalent ligand such as the rigidity of the scaffold on which the carbohydrate determinants are presented and the orientation of each recognition element.

Although the chelate effect often has been invoked to explain enhancements seen for the functional affinities of multivalent saccharide derivatives, to span the large distances that separate many carbohydrate binding sites (e.g. $\geq 20 \, \alpha$) requires the payment of a greater conformational entropy penalty. Interestingly, many multivalent ligands designed to bind to separate subsites on a multimeric protein seem to function by clustering their target proteins at the cell surface rather than direct occupation of the binding sites within the multimer (Fig. 4B) (Glick and Knowles 1991; Glick et al. 1991). This binding mode can be viewed as a manifestation of the chelate effect. The translational entropy penalties of binding a membrane-bound receptor are not as great as those incurred for receptor–ligand interactions in solution, because membrane localization of receptors decreases their degrees of freedom. Membrane-tethered receptors can diffuse in the bilayer, and several different geometries may be available that can accommodate multipoint binding. Similarly, the chelate effect may be important for analysis of binding to polymeric or polyvalent displays. In many of these cases, there are a large number of states that allow occupation of more than one saccharide-binding site, leading to increased probabilities that multivalent binding will occur. Non-natural multivalent arrays can be used to evaluate the importance of the chelate effect and to investigate the mechanisms underlying increases in functional affinities.

10.5 Statistical Effects of High Binding Site Density

The display of either a protein receptor and/or its cognate ligand at a high density can lead to an increase in the measured affinity of interaction, even if multipoint binding does not occur. In many physiologically relevant settings, such as on the cell surface, this mode of substrate presentation commonly occurs. The increase in apparent affinity that results from such a display, in which a high effective concentration of one or both binding partners is achieved, can be rationalized by modeling the binding event as a two step process (Berzofsky et al. 1993). In the case of a multivalent ligand, the receptor can form an encounter complex with the ligand in the first step. In the second stage, the stable complex is generated. The probability that the encounter complex $(R...L)$ will go on to form the final complex will be higher when the ligand is multivalent than when it is monovalent; the frequency with which the receptor diffuses away from the ligand will be diminished because it is more likely to arrive at a productive binding mode with such a ligand (Fig. 4C). For example, in Fig. 4C, the multivalent saccharide derivative can form two receptor-ligand interactions (R_1-L_1 and R_2-L_2). Although saccharide epitopes L_3 and L_4 are not directly participating in the binding interaction, their proximity increases the likelihood that the multivalent ligand will rebind to the receptors rather than diffusing away. This model has been invoked to explain the enhanced lectin affinities of some oligosaccharides relative to monosaccharide analogs (Mandal et al. 1994). The decrease in dissociation rates for low molecular weight oligosaccharides (3–6 saccharide residues) are likely to be quite modest (affinity increases of two to ten-fold).

10.6 Extended and Secondary Binding Sites

Some multivalent ligands may bind more tightly to their protein targets because additional saccharide recognition elements outside the primary site can occupy extended binding sites. In such cases, the residues that interact with the extended region can contribute to complexation by a different means than the recognition determinant that binds in the primary site. For example, two proteins involved in innate immunity, the mannose-binding protein A (MBP-A) and the mannose-binding protein C (MBP-C), have primary binding sites with similar saccharide selectivities, yet only the MBP-C has an extended cleft. Because of this divergence in binding site attributes (Lee and Lee 1997), MBP-C accommodates linear mannose oligosaccharides but serum MBP-A, which binds cell surfaces, does not. In another example, the structure of the plant lectin from *Lathyrus ochrus* (LOL II) complexed to a decasaccharide from human lactotransferrin contains extended interactions between oligosaccharide and the protein (Bourne et al. 1994). In this complex, a mannose residue occupies the primary binding site and its oligosaccharide scaffold is oriented to place a fucose branch into a cleft adjacent to the primary site. Other protein–saccharide interactions similarly benefit from extended binding sites, with some saccharides gaining additional binding energy through indirect, water-mediated contacts to the protein surface (Weis and Drickamer 1996). Although a detailed understanding of the role of solvation in protein-saccharide interactions is lacking (Toone 1994), the attachment of various functional groups to regions of the saccharide that do not directly contact the binding site can either diminish or augment the affinity of interaction (Lemieux 1996; Lemieux et al. 1994). Multidentate saccharide derivatives, therefore, can benefit from van der Waals contacts to protein regions adjacent to the main binding site and by altering solvation around the saccharide determinant. With these potential options, the strength of binding for saccharide residues presented on various scaffolds can vary even when the scaffold itself does not appear to participate in any direct interactions. Although multivalent ligands that can occupy these extended binding sites will bind more tightly than their monovalent counterparts, the affinity increases generally will be smaller than those that are observed when the multiple receptor–ligand complexes form.

Structural studies of some carbohydrate-binding proteins have revealed secondary saccharide binding sites that are structurally distinct from the primary binding sites (Fig. 4D) (Hester et al. 1995; Hester and Wright 1996; Wright 1984). For example, secondary binding regions have been observed for influenza virus hemagglutinin (Sauter et al. 1992) and for MBP-C (Ng et al. 1996). In both cases, the secondary sites bind with lower affinities to saccharide ligands than does the primary. Although the biological relevance of such sites remains obscure, multivalent ligands with the ability to occupy both sites simultaneously will be more effective inhibitors. In such systems, however, apparent affinity increases derived from multivalent presentation may not be as dramatic as those observed for the formation of multiple identical receptor-ligand complexes. A secondary saccharide binding site may have different selectivity than the primary, as is the case for MBP-C (Ng et al. 1996; Quesenberry et al. 1997). Consequently, in addition to their ability to augment the affinity of binding for selected oligosaccharides, secondary binding sites endow lectins with the ability to discriminate between multivalent ligands.

10.7 Steric Effects

Whitesides and coworkers gathered support for steric stabilization, another mechanism that contributes to multivalent ligand potency. In their studies of inhibition of influenza virus-mediated cell agglutination (Mammen et al. 1995; Sigal et al. 1996) using sialic acid-substituted acrylamide polymers, they found that the effectiveness of the polymeric materials at preventing the virus-erythrocyte interaction may be due to more than just the chelate effect. In this system the viral hemagglutinin binds to sialic acid residues on the red blood cell surface, causing the cells to agglutinate. Once the polymers begin to interact with the viral surface, their large size compounded by their large hydration shell can sterically obstruct viral interactions with the red blood cell surface. Circumstantial evidence suggests cells may use similar mechanisms to avoid adhesion in vivo (Fryer and Hockfield 1996; Manjunath et al. 1995; Rutishauser et al. 1988).

Steric stabilization by the polymer chain requires binding to the viral surface, but high occupancy of sialic acid binding sites is neither necessary nor desirable. For example, when a monovalent sialic acid analog is added along with the polymers, greater inhibition is observed (Choi et al. 1996). This enhancement is attributed to increased steric stabilization brought about by the monovalent compound displacing some of the sialic acid residues linked to the polymer. By this process, the polymer backbone would be rendered a more effective blockade (Mammen et al. 1995; Sigal et al. 1996). These studies suggest a mechanism by which steric effects could play some role in the inhibition of agglutination, yet enhanced avidity for the virus due to specific ligand binding appears to be a major mechanism of inhibition in these systems as well.

10.8 Synthetic Scaffolds for Multivalent Presentation

The variety of mechanisms by which multivalent presentation of saccharide determinants can result in increased functional affinities presents new challenges. Developing an understanding of monovalent protein-saccharide interactions would be a great step forward, but it alone is not enough. In general, studies of multivalent protein-saccharide interactions involve systems that preclude precise dissection of the energetic contributions to binding. The challenge is to develop methods to elucidate the potential mechanisms that contribute to these interactions. Features pertinent to multipoint binding include the intrinsic affinity of the monovalent protein-saccharide interaction, the number of recognition elements displayed, the distance and orientation of these elements, and the conformational flexibility of the scaffold on which these binding units are displayed. The development of methods to address these issues is critical to our understanding of a host of physiologically and medically significant multivalent binding events. Chemical synthesis is providing the means to address the complex recognition questions in this area. Our applications of synthetic multivalent saccharide derivatives are outlined below.

10.9 Structure/Function Studies of Neoglycopolymer Inhibitors of Concanavalin A-Mediated Cell Agglutination

The demonstration that saccharide-substituted polymers (neoglycopolymers) created by ROMP were potent inhibitors of saccharide-binding proteins led us to examine the specificity of multivalent binding. Three issues that can be addressed by changing the ligand structure are: (1) the effect of altering the recognition elements in a polyvalent array, (2) the role of carbohydrate residue density in multivalent binding, and (3) the number of residues presented in the multivalent array. Alternatively, the affinities of ligands for different saccharide-binding proteins are also of interest. We have probed these issues with our multivalent, glycoprotein-like materials generated by ROMP.

10.10 Multivalent Ligand Specificity: Structure of the Recognition Elements

An important question in carbohydrate recognition is: can low affinity interactions mediate specific binding? We hypothesized that if the chelate effect was important in the recognition of multidentate saccharide ligands, differences between monovalent ligands might be amplified when they are presented in a multivalent array (Fig. 1). To test these ideas, we compared the interactions of concanavalin A with monovalent and multivalent mannose and glucose derivatives (Fig. 5) (Mortell et al. 1996). We found that the activities of the monovalent α-linked C-glycosides were very similar in the cell agglutination assay (Fig. 5), although the C-mannoside derivative bound with slightly higher affinity (1.5-fold) than the corresponding glucoside in a direct binding assay (Weatherman et al. 1996). Significantly, the multivalent ligands were more selective than were the corresponding monovalent compounds. For instance, the C-mannose-substituted polymer **3** was 100-fold more effective on a saccharide residue basis than its C-glucoside counterpart **2**. Since the monomeric α-C-linked sugars exhibited identical activities, this result highlights the greater selectivity multivalency can impart. Moreover, the 4-fold difference between the monovalent glucose and mannose O-glycosides is also amplified. When the activities of the multivalent O-glycosides were compared, an even greater difference

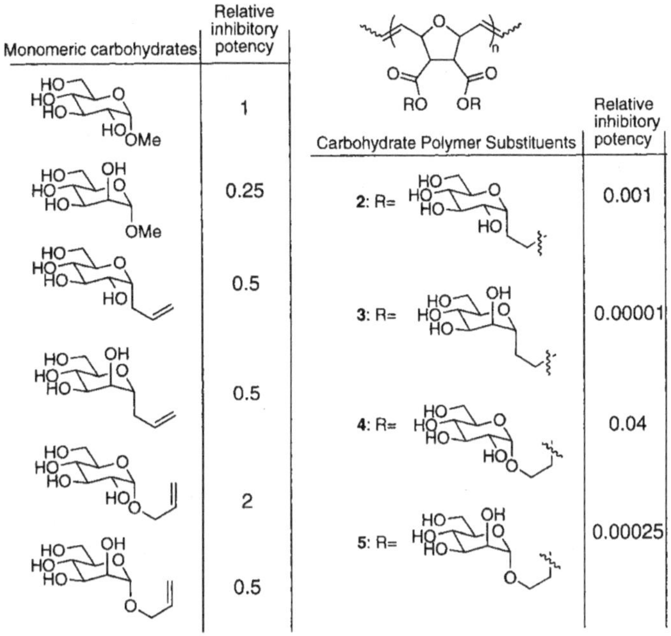

Fig. 5. Structure and biological activities of carbohydrate monomers and carbohydrate-substituted polymers generated by ROMP. Inhibitory potency is defined as: the saccharide residue concentration needed to inhibit the agglutination of red blood cells mediated by the protein concanavalin A. Values have been standardized relative to monovalent α-methyl glucopyranoside

(160-fold) was found between the multivalent O-glucoside and O-mannoside ligands, compounds **4** and **5**, respectively. The magnitude of these differences is greater than expected by simple models based on the chelate effect.

Subsequent to our studies, another example of increased specificity through multivalent interactions was reported (Liang et al. 1996, 1997). These workers generated a combinatorial library on Tentagel beads, which they then screened for its ability to interact with a lectin. They found high selectivities, but these did not correlate with intrinsic solution affinities of the monovalent saccharide derivatives for the protein.

The library, however, contained non-natural saccharides modified with aromatic functional groups, which were presented on Tentagel beads. This feature renders interpretation of the binding data quite compli-cated, because many lectins have subsites that interact with aromatic groups (Lis and Sharon 1986). Thus, the physiological relevance of investigations of multivalency employing recognition epitopes with such groups is unclear. Still, these studies, like those described by us, point to the importance of considering other mechanisms in addition to the chelate effect when analyzing multivalent recognition events.

10.11 Multivalent Synthetic Ligands Differing in Average Lengths

The observed enhancement of specificity for multivalent binding by concanavalin A could be consistent with the chelate effect, but we sought to gather additional information about the mechanism of en-hancement. The affinity, specificity and kinetics of recognition events mediated by multivalent protein-carbohydrate complexes will depend on the size of the multidentate ligand and the number of recognition elements it presents. To investigate this issue, we developed ROMP for the syntheses of saccharide-substituted oligomers of varying lengths (Kanai et al. 1997).

ROMP catalyzed by well-defined metal carbene complexes provides opportunities for generating unique scaffolds for multivalent ligand display (Kiessling and Strong). In a living polymerization, elongation proceeds without termination or chain transfer processes, and ROMP can be a living process (Ivin and Mol 1997). A living polymerization that proceeds through a rapid chain initiation and slower elongation steps can be used to synthesize materials of different lengths by control-ling the ratio of initiator to monomer (Lynn et al. 1996). Studies of the mechanism of ROMP by the Grubbs group (Fig. 6) have revealed that the defined ruthenium carbene **6** (Fig. 7; Schwab et al. 1995) affords initiation rates with reactive monomers that are approximately 9-fold faster than elongation rates (Dias et al. 1997). Thus, ROMP has the potential to be an excellent solution to the problem of generating librar-ies of oligomers composed of different length chains.

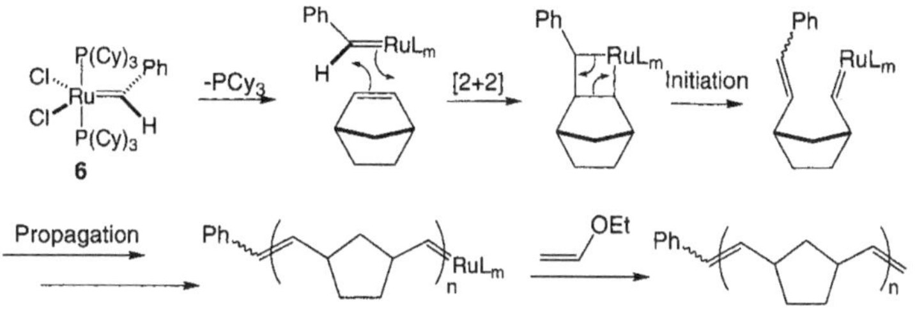

Fig. 6. An overview of the ring-opening metathesis polymerization initiated by defined ruthenium complexes

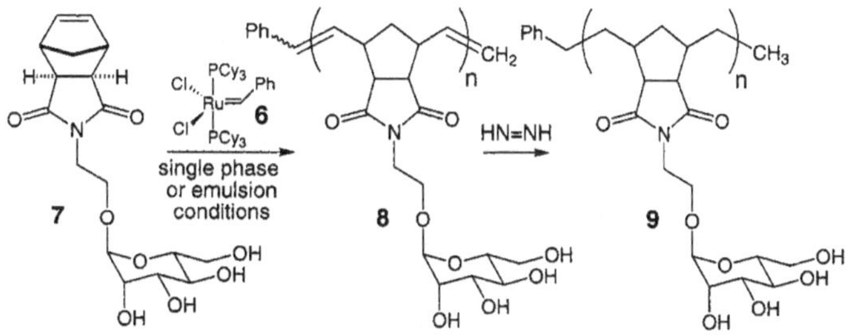

Fig. 7. A living polymerization of mannose-substituted norbornene derivative was used to produce materials of defined lengths for biological testing. Diimide reduction of the alkene backbone affords saturated polymers **9** for comparison with unsaturated derivatives **8**

ROMP was used to create a series of multivalent mannose derivatives that varied in the average length. Because living polymerizations rely on the propensity of the monomer to engage in elongation, template **7** (Fig. 7) was devised to undergo more rapid metathesis than the carbohydrate-substituted substrates previously employed. The increased reactivity was envisioned to arise from two features. First, the strain in the bicyclo[2.2.1] system is augmented by fusion to the cyclic imide, which should increase initiation and propagation rates. Second, the more electron rich norbornene derivative was selected over the 7-oxanorbornene analog to favor elongation. The desired monomer was generated by

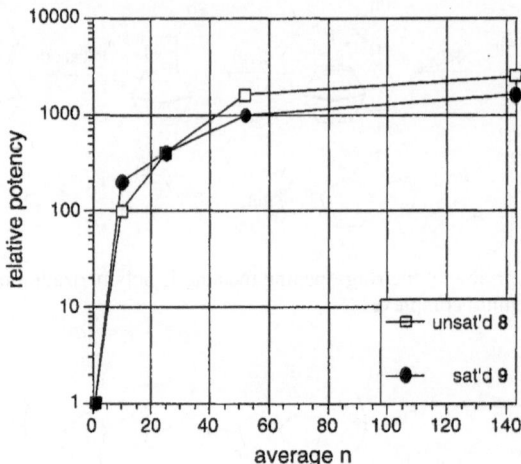

Fig. 8. Relative inhibitory potencies of neoglycopolymers of different lengths **8** (unsaturated) and **9** (saturated) in a concanavalin A-mediated cell agglutination assay (Fig. 7). Potencies are reported on a saccharide residue basis and standardized to the monovalent α-methyl mannopyranoside. The average lengths (n = average number of residues) of the polymers were determined by ^{1}H NMR integration and verified by gel permeation chromatography

attachment of an α-linked mannose residue to an imide-containing norbornene scaffold.

The target compounds were assembled using the defined ruthenium carbene **6**, using either single phase or emulsion polymerization conditions (Fig. 7). The oligomerization of monomer was effected with ruthenium carbene, using increasing monomer-to-catalyst ratios (m/c) to produce polymers of increasing length (Fig. 7). By altering reaction conditions, a series of neoglycopolymers of increasing degree of polymerization (average length) could readily be obtained. The polymerization reactions displayed the expected characteristics of living polymerizations (a linear dependence on the degree of polymerization relative to the m/c ratio). Moreover, the resulting materials, **8**, had narrow polydispersity indices (≤ 1.15) as determined by gel filtration chromatography (Kanai et al. 1997). To provide insight into the impact of polymer backbone flexibility on activity, the components of set 8 were reduced to produce a parallel series of neoglycopolymers with saturated backbones (Fig. 7, series **9**).

In assays with concanavalin A, the potencies of these materials increased dramatically with increasing polymer length (Fig. 8). Materials composed of 50 residues or more had the highest efficacies (approximately 2000-fold more active on a saccharide residue basis). Analysis of the concanavalin A x-ray structure reveals that the distance between saccharide binding sites within the tetramer is approximately 65 Å (Fig. 3) (Derewenda et al. 1989; Hamodrakas et al. 1997; Loris et al. 1996). Molecular modeling (Mohamadi et al. 1990) studies of the neoglycopolymer indicate that materials composed of ≥35 saccharide residues can span two binding sites within the tetramer. A mixture of stereoisomers is produced in the polymerization reactions, so we focused on the limiting cases (Fig. 9). Specifically, the most compact structures derived from the formation of cis alkene backbone isomers require approximately 35 monomer units to span the saccharide binding sites. Alternatively, polymers with backbones of the all trans configuration composed of 17 units could reach across two saccharide binding sites (Fig. 9). It is therefore interesting to note that the activity of the polymers increases dramatically until almost all constituents of the population have the ability to place saccharide residues in two sites (i.e., average length of 50).

The variation in potency between polymers of different average length is consistent with contributions from the chelate effect (Fig. 4A). In this model, a single polymer chain of the proper length interacts simultaneously with two binding sites on the protein; only a single translational entropy penalty is paid, while favorable enthalpic contributions from both sites will contribute to binding. Consistent with this idea, materials that can place saccharide residues simultaneously in two recognition sites of the concanavalin A tetramer are more efficacious than those that can only occupy a single binding site.

The chelate effect alone, however, cannot explain the data. Oligomers composed of more residues than the number necessary to span the sites should display decreased activities when evaluated on a saccharide residue basis, but they do not. Consequently, the potency of the longer polymers is greater than chelation predicts. The efficacies of these longer polymers and of oligomers too short to span the protein binding sites can be attributed to a high local concentration of mannose residues,

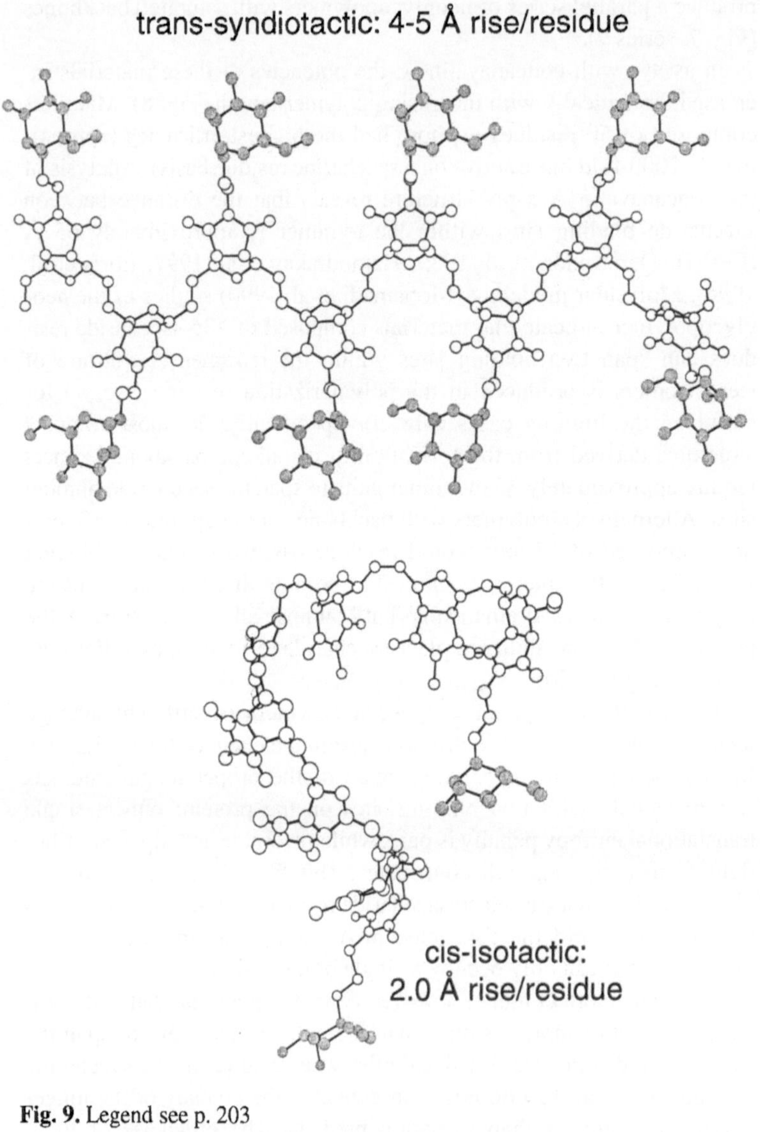

trans-syndiotactic: 4-5 Å rise/residue

cis-isotactic:
2.0 Å rise/residue

Fig. 9. Legend see p. 203

which would perturb the rate of dissociation of multivalent ligands. Neoglycopolymers that contain more saccharide recognition elements will exhibit slower dissociation rates, with rebinding becoming more favorable and dissociation less so. Moreover, the large number of mannose residues available for binding with these ligands suggests that a number of different complexes can exist. Thus, longer multivalent ligands should have a higher probability of interacting with two binding sites in comparison to that for divalent ligands. Finally, steric stabilization could also contribute to neoglycopolymer stability. For example, the binding of concanavalin A to the larger neoglycopolymers may sterically occlude other saccharide binding sites within the tetramer.

10.12 Recognition of Multivalent Ligands by Different Saccharide-Binding Proteins

Our findings indicate that the chelate and statistical effects are important in the recognition of multivalent saccharide derivatives by concanavalin A. We wished to determine the generality of our observations by further investigating mechanisms of multivalent binding with other lectins (Weatherman et al. 1996). To ascertain whether other saccharide-binding proteins exhibit similar properties, our multivalent ligands of differing lengths were tested for their abilities to inhibit cell agglutination by another mannose-binding protein of known structure, the snowdrop lectin (Hester et al. 1995; Hester and Wright 1996).

The quaternary structures of concanavalin A and the snow drop lectin differ markedly; consequently, the spatial dispositions of their saccharide-binding clefts are also distinct. Although both proteins are tetramers, the snowdrop lectin possesses a total of 12 mannose-binding sites, a much larger number than the 4 displayed by concanavalin A. Addition-

Fig. 9. Molecular models generated by molecular mechanics calculations (Macromodel Modeling Package) for two regular structures of mannose-substituted oligomers derived from ROMP (see Fig. 7, compound 8). Polymers composed of all trans alkenes present saccharide residues at distances of approximately 4–5 Å/residue. Materials derived from all cis alkenes are more compact, with a 2 Å rise/residue. The neoglycopolymers used in the binding studies were derived from a mixture of cis and trans backbone isomers

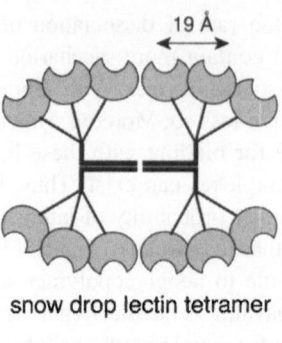

snow drop lectin tetramer

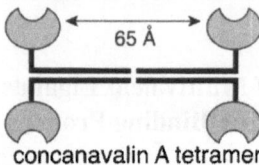

concanavalin A tetramer

Fig. 10. The concanavalin A and snowdrop lectin tetramers. Concanavalin A presents 4 saccharide binding sites while the snowdrop lectin has 12

ally, the separation between sites within the snowdrop lectin is 19 Å (Hester et al. 1995; Hester and Wright 1996), which is considerably closer than the 65 Å distance in the concanavalin A tetramer (Derewenda et al. 1989) (Fig. 10). If the snowdrop lectin bound monovalent mannose residues with the same intrinsic affinity as concanavalin A, it could exhibit very high functional affinities for multivalent mannose derivatives. This is not the case: the minimum dose for inhibition of agglutination by α-methyl mannopyranoside is 3.2 mM for concanavalin A but 400 mM for the snowdrop lectin (Kanai and Kiessling, unpublished results). It is intriguing to note that the high number of binding sites and their short separation distance within the snowdrop lectin structure are offset by the weak affinity of the protein for monovalent mannose derivatives. Perhaps natural systems have evolved to balance saccharide affinity and multipoint binding. The balance could be essential for the maintenance of selective interactions through multivalent binding.

The greater number of binding sites within snowdrop lectin could result in the multivalent ligands exhibiting extremely high increases in

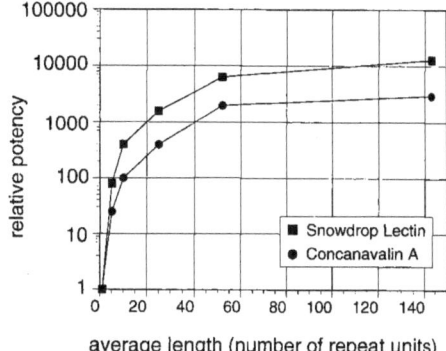

average length (number of repeat units)

Fig. 11. Relative inhibitory potencies of saturated neoglycopolymers 9 (Fig. 7) of different lengths in cell agglutination assays using either concanavalin A or the snowdrop lectin. Potencies are reported on a saccharide residue basis and standardized to the monovalent α-methyl mannopyranoside. The average lengths (average number of residues) of the polymers were determined by ^{1}H NMR integration. The data reveal the marked increases in potency afforded by the multivalent ligands

potency. Consequently, we tested our series of polymers composed of different average lengths in a snowdrop lectin-facilitated cell agglutination assay for comparison with the results obtained with concanavalin A (Fig. 7). Overall, dramatic increases in activities were observed for longer neoglycopolymers (Fig. 11). These results highlight the important role of multivalency in recognition by both the snowdrop lectin and concanavalin A.

The agglutination data from the two lectins provide additional evidence that the chelate effect contributes to multivalent recognition by saccharide-binding proteins. Analysis of the polymer models suggests that oligomers composed of only 4–10 monomer units should be able to span the snowdrop lectin binding sites (Fig. 9). As predicted from these estimates, shorter oligomers (5 and 10 residues long) exhibit greater enhancements in potency when tested with snowdrop lectin than with concanavalin A. These data highlight the advantages of multivalent binding for proteins with low intrinsic saccharide-binding abilities.

The maximum increase in ligand potency for snowdrop lectin is greater than that for concanavalin A. The most active inhibitors of concanavalin A-mediated agglutination are 2000-fold more effective on a saccharide residue basis than the corresponding monovalent mannose derivatives, but increases up to 13000-fold are obtained for snowdrop lectin. This observation is notable as the snowdrop lectin has such weak affinity for monovalent ligands that one might assume that the conformational entropy penalty for occupying additional carbohydrate binding sites could outweigh any benefit of multipoint binding. It is interesting to note, however, that this system appears to bind avidly to multivalent ligands. The dependence of oligomer length on activity is consistent with the type of enhancements predicted by the chelate effect, although other mechanisms, such as those described previously, likely contribute as well. Investigations which reveal the similarities between saccharide binding by concanavalin A and the snowdrop lectin provide further support for the importance of multivalent interactions in protein–saccharide interactions and insight into the mechanisms by which low affinity interactions function in biological systems.

10.13 Conclusions

Although it is not well understood, multivalent binding is critical in physiological settings. Cell adhesion, antibody–antigen recognition, and some signal transduction events rely on multipoint receptor–ligand interactions to produce the proper response. We have found that emerging techniques in organometallic and polymer chemistry can be used to study these complex, yet fundamental problems in biomolecular recognition. Our studies indicate that several factors contribute to the high functional affinities often observed in multivalent recognition events. For example, oligomeric or membrane-associated proteins that bind natural multivalent saccharide displays may benefit from the high local concentration of recognition epitopes presented through glycoprotein or glycolipid clusters (Fig. 12). In these configurations, both the effects of decreased off-rates and multipoint binding can contribute to the functional affinities observed in the interactions. Additional investigations will illuminate further the molecular mechanisms that give rise to affinity and specificity in such systems.

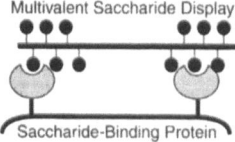

Multivalent Saccharide Display

Saccharide-Binding Protein

1. Glycolipids
2. N-Glycoproteins
   Branched Saccharide Epitopes
3. O-Glycoproteins
   Mucins Have Highly Clustered
   Glycosylated Ser/Thr Residues

Fig. 12. Carbohydrate-binding proteins in physiological settings appear to take advantage of the multivalent presentation of saccharides that occurs in glycoproteins and glycolipid displays. One mechanism by which increased functional affinities for such recognition events can be achieved is through multipoint binding facilitated by the high local concentrations of saccharide residues

References

Bell GI (1974) Model for the binding of multivalent antigen to cells. Nature 248:430–431

Berzofsky JA, Berkower IJ, Epstein SL (1993) Antigen–antibody interactions and monoclonal antibodies. In: Paul WE (ed) Fundamental immunology. Raven, New York, pp 421–465

Bittiger H, Schnebli HP (1976) Concanavalin A as a Tool. Wiley, London, pp 639

Boullanger P (1997) Amphiphilic carbohydrates as a tool for molecular recognition in organized systems. Topics Curr Chem 187:275–312

Bourne Y, Mazurier J, Legrand D, Rougé P, Montreuil J, Spik G, Cambillau C (1994) Structures of a legume lectin complexed with the human lactotransferrin N2 fragment, and with an isolated biantennary glycopeptide: role of the fucose moiety. Structure 2:209–19

Bovin NV, Gabius H-J (1995) Polymer-immobilized carbohydrate ligands: versatile chemical tools for biochemistry and medical sciences. Chem Soc Rev 24:413–421

Choi S-K, Mammen M, Whitesides GM (1996) Monomeric inhibitors of influenza neuraminidase enhance the hemagglutination inhibition activities of polyacrylamides presenting multiple C-sialoside groups. Chem Biol 3:97–104

Crothers DM, Metzger H (1972) The influence of polyvalency on the binding properties of antibodies. Immunochemistry 9:341–57

Derewenda Z, Yariv J, Helliwell JR, Kalb AJ, Dodson EJ, Papiz MZ, Wan T, Campbell J (1989) The structure of the saccharide-binding site of concanavalin A. EMBO J 8:2189–2193

Dias EL, Nguyen ST, Grubbs RH (1997) Well-defined ruthenium olefin metathesis catalysts: mechanism and activity. J Am Chem Soc 119:3887–3897

Drickamer K (1995) Multiplicity of lectin-carbohydrate interactions. Struct Biol 2:437–439

Dwek RA (1996) Glycobiology: toward understanding the function of sugars. Chem Rev 96:683–720

Elgavish S, Shaanan B (1997) Lectin-carbohydrate interactions: different folds, common recognition principles. Trends Biochem Sci 22:462–467

Feizi T (1993) Oligosaccharides that mediate mammalian cell-cell adhesion. Curr Opin Struct Biol 3:701–710

Feizi T, Stoll MS, Yuen C-T, Chai W, Lawson AM (1994) Neoglycolipids: probes of oligosaccharide structure, antigenicity and function. Methods Enzymol 230:484–519

Fraser C, Grubbs RH (1995) Synthesis of glycopolymers of controlled molecular weight by ring opening metathesis polymerization using well-defined ruthenium carbene catalysts. Macromolecules 28:7248–7255

Fryer HJL, Hockfield S (1996) The role of polysialic acid and other carbohydrate polymers in neural structural plasticity. Curr Opin Neurobiol 6:113–118

Gabius H-J (1997) Animal lectins. Eur J Biochem 243:543–576

Glaudemans CPJ (1991) Mapping of subsites of monoclonal, anti-carbohydrate antibodies using deoxy and deoxyfluoro sugars. Chem Rev 91:25–33

Glick GD, Knowles JR (1991) Molecular recognition of bivalent sialosides by influenza virus. J Am Chem Soc 113:4701–4703

Glick GD, Toogood PL, Wiley DC, Skehel JJ, Knowles JR (1991) Ligand recognition by influenza virus. J Biol Chem 266:23660–23669

Hamodrakas SJ, Kenellopoulos PN, Pavlou K, Tucker PA (1997) The crystal structure of the complex of concanavalin A with 4'-methylumbelliferyl-α-D-glucopyranoside. J Struct Biol 118:23–30

Hansen HC, Haataja S, Finne J, Magnusson G (1997) Di-, tri-, and tetravalent dendritic galabiosides that inhibit hemagglutination by Streptococcus suis at nanomolar concentration. J Am Chem Soc 119:6974–6979

Hester G, Wright CS (1996) The mannose-specific bulb lectin from Galanthus nivalis (snowdrop) binds mono- and dimannosides at distinct sites. Structure analysis of refined complexes at 2.3 angstrom and 3.0 angstrom resolution. J Mol Biol 262:516–531

Hester G, Kaky H, Goldstein IJ, Wright CS (1995) Structure of mannose-specific snowdrop (galanthus-nivalus) lectin is representative of a new plant lectin family. Nat Struct Biol 2:472–479

Ivin KJ, Mol JC (1997) Olefin metathesis and metathesis polymerization. Academic, San Diego

Jencks WP (1981) On the attribution and additivity of binding energies. Proc Natl Acad Sci USA 78:4046–4050

Kanai M, Mortell KH, Kiessling LL (1997) Varying the size of multivalent ligands: the dependence of concanavalin A inhibition on neoglycopolymer length. J Am Chem Soc 119:9931–9932

Kiessling LL, Pohl NL (1996) Strength in numbers: non-natural polyvalent carbohydrate derivatives. Chem Biol 3:71–77

Kiessling LL, Strong LE (1998) Bioactive polymers. In: Fürstner A (ed) Alkene metathesis in organic synthesis. Springer, Berlin Heidelberg New York (Topics in organometallic chemistry, vol 2) (in press)

Laine RA (1994) A calculation of all possible oligosaccharide isomers both branched and linear yields 1.05×10^{12} structures for a reducing hexasaccharide: the isomer barrier to development of single-method saccharide sequencing or synthesis systems. Glycobiology 4:759–767

Lee RT, Lee YC (1987) Preparation of cluster glycosides of N-acetyl galactosamine that have subnanomolar binding constants towards the mammalian hepatic Gal/GalNAc specific receptor. Glycoconjugate J 4:317–328

Lee RT, Lee YC (1997) Difference in binding-site architecture of the serum-type and liver-type mannose-binding proteins. Glycoconjugate J 14:357–363

Lee YC, Lee RT (1995) Carbohydrate-protein interactions: basis of glycobiology. Acc Chem Res 28:321–327

Lemieux RU (1996) How water provides the impetus for molecular recognition in aqueous solution. Acc Chem Res 29:373–380

Lemieux RU, Du M-H, Spohr U (1994) Relative effects of ionic and neutral substituents on the binding of an oligosaccharide by a protein. J Am Chem Soc 116:9803–9804

Liang R, Yan L, Loebach J, Ge M, Uozumi Y, Sekanina K, Horan N, Gildersleeve J, Thompson C, Smith A, Biswas K, Still WC, Kahne D (1996) Parallel synthesis and screening of a solid phase carbohydrate library. Science 274:1520–1522

Liang R, Loebach J, Horan N, Ge M, Thompson C, Yan L, Kahne D (1997) Polyvalent binding to carbohydrates immobilized on an insoluble resin. Proc Natl Acad Sci USA 94:10554–10559

Lis H, Sharon N (1986) Lectins as molecules and as tools. Annu Rev Biochem 55:35–67

Lis H, Sharon N (1998) Lectins: carbohydrate-specific proteins that mediate cellular recognition. Chem Rev 98:637–674

Loris R, Maes D, Poortmans F, Wyns L, Bouckaert J (1996) A Structure of the complex between concanavalin A and methyl-3,6-di-O-(α-D-mannopyrano-syl)α-D-mannopyranoside reveals two binding modes. J Biol Chem 271:30614–30618

Lynn DM, Kanaoka S, Grubbs RH (1996) Living ring-opening metathesis polymerization in aqueous media catalyzed by well-defined ruthenium carbene complexes. J Am Chem Soc 118:784–790

Mammen M, Dahmann G, Whitesides GM (1995) Effective inhibitors of hemagglutination by influenza virus synthesized from polymers having active ester groups. Insight into the mechanism of inhibition. J Med Chem 38:4179–4190

Mandal DK, Kishore N, Brewer CF (1994) Thermodynamics of lectin-carbohydrate interactions. Titration microcalorimetry measurements of the binding of N-linked carbohydrates and ovalbumin to concanavalin A. Biochemistry 33:1149–1156

Manjunath NC, M., Ardman M, Ardman B (1995) Negative regulation of T-cell adhesion and activation by CD43. Nature 377:535–538

Mohamadi F, Richards NGJ, Guida WC, Liskamp R, Lipton M, Caufield C, Chang C, Hendrickson T, Still WC (1990) Macromodel – an integrated software system for modeling organic and bioorganic molecules using molecular mechanics. J Comput Chem 11:440–67

Mortell KH, Gingras M, Kiessling LL (1994) Synthesis of cell agglutination inhibitors by ring–opening metathesis polymerization. J Am Chem Soc 116:10293–10294

Mortell KH, Weatherman RV, Kiessling LL (1996) Recognition specificity of neoglycopolymers prepared by ring-opening metathesis polymerization. J Am Chem Soc 118:2297–2298

Ng KK-S, Drickamer K, Weis WI (1996) Structural analysis of monosaccharide recognition by rat liver mannose-binding protein. J Biol Chem 271:663–674

Nomura K, Schrock RR (1996) Preparation of "sugar coated" homopolymers and multiblock ROMP copolymers. Macromolecules 29:540–545

Osawa T, Matsumoto I (1972) Gorse (Ulex europeus) phytohemagglutinins. Methods Enzymol 28:323–327

Page MI, Jencks WP (1971) Entropic contributions to rate accelerations in enzymic and intramolecular reactions and the chelate effect. Proc Natl Acad Sci USA 68:1678–1683

Quesenberry MS, Lee RT, Lee YC (1997) Difference in the binding mode of two mannose-binding proteins: Demonstration of a selective minicluster effect. Biochemistry 36:2724–2732

Quiocho FA, Vyas NK, Spurlino JC (1989) Atomic interactions between proteins and carbohydrates. Transactions ACA 25:23–35

Rao JH, Lahiri J, Isaacs L, Weis RM, Whitesides GM (1998) A trivalent system from vancomycinD-Ala-D-Ala with higher affinity than avidinbiotin. Science 280:708–711

Rini JM (1995) Lectin structure. Annu Rev Biophys Biomol Struct 24:551–577

Roy R (1996) Syntheses and some applications of chemically defined multivalent glycoconjugates. Curr Opin Struct Biol 6:692–702

Roy R, Park WKC, Srivastava OP, Foxall C (1996) Combined glycomimetic and multivalent strategies for the design of potent selectin antagonists. Bioorg Med Chem Lett 6:1399–1402

Rutishauser U, Acheson A, Hall AK, Mann DM, Sunshine J (1988) The neural cell adhesion molecule (NCAM) as a regulator of cell-cell interactions. Science 240:53–57

Sauter NK, Glick GD, Crowther RL, Park S-J, Eisen MB, Skehel JJ, Knowles JR, Wiley DC (1992) Crystallographic detection of a second ligand binding site in influenza virus hemagglutinin. Proc Natl Acad Sci USA 89:324–328

Schnarr RL, Weigel PH, Kuhlenschmidt MS, Lee YC, Roseman S (1979) Adhesion of chicken hepatocytes to polyacrylamide gels derivatized with N-acetylglucosamine. J Biol Chem 253:7940–7951

Schwab P, France MB, Ziller JW, Grubbs RH (1995) A series of well-defined metathesis catalysts–synthesis of $[RuCl_2(=CHR')(PR_3)_2]$ and its reactions. Angew Chem Int Ed Engl 34:2039–41

Sigal GB, Mammen M, Dahmann G, Whitesides GM (1996) Polyacrylamides bearing pendant α-sialoside groups strongly inhibit agglutination of erythrocytes by influenza virus: the strong inhibition reflects enhanced binding through cooperative polyvalent interactions. J Am Chem Soc 118:3789–3800

Spaltenstein A, Whitesides GM (1991) Polyacrylamides bearing pendant α-sialoside groups strongly inhibit agglutination of erythrocytes by influenza virus. J Am Chem Soc 113:686–687

Toone (1994) Structure and energetics of protein-carbohydrate complexes. Curr Opin Struct Biol 4:719–728

Varki A (1993) Biological roles of oligosaccharides: all of the theories are correct. Glycobiology 3:97–130

Weatherman RV, Mortell KH, Chervenak M, Kiessling LL, Toone EJ (1996) Specificity of C-glycoside complexation by mannose/glucose specific lectins. Biochemistry 35:3619–3624

Weis WI, Drickamer K (1996) Structural basis of lectin-carbohydrate recognition. Annu Rev Biochem 65:441–473

Wong SYC (1995) Neoglycoconjugates and their applications in glycobiology. Curr Opin Struct Biol 5:599–604

Wright CS (1984) Structural comparison of the two distinct sugar binding sites in wheat germ agglutinin isolectin II. J Mol Biol 178:91–104

11 Self-Organized Autocatalytic Chemical Networks and Molecular Ecosystems: Do They Provide the Experimental Tools for Modeling the Transition from Inanimate to Animate Chemistry?

M.R. Ghadiri

11.1   Introduction 213
11.2   Molecular Self-Replication 217
11.3   De Novo Design of Peptide Catalysts 219
11.4   Design of a Synthetic Peptide Ligase 220
11.5   Design of Self-Replicating Peptides 223
11.6   Self-Organized Autocatalytic Networks 229
11.7   Reciprocal Networks 229
11.8   Autocratic and Parasitic Networks 231
11.9   Hypercyclic Networks 232
References 235

11.1 Introduction

Despite remarkable advances in understanding the molecular basis of life, little is known about how in a living system the undeniably inanimate chemical transformations give rise to its animate characteristics. While chemists have long sought to design biomimetic model systems to better understand the structural and functional underpinning of biomolecules, only recently have chemical model systems started to become available that may be used to address some of the fundamental processes which biomolecules exploit to self-organize into animate systems (Lee et al. 1997). As a guide for an experimental program toward this issue, a working concept is presented that takes into account recent advances in the mathematical understanding of complex nonlinear systems, theories on system self-organization, and new discoveries from this and other laboratories on the design of self-organized autocatalytic networks. Specifically, the design, study, and discovery of peptide model systems are highlighted that can mimic the initial steps of the transition from simple molecular mixtures into molecular ecosystems: namely the onset of self-organization and the appearance of the phenomenon of emergence.

A living system is an autonomous self-sustained chemical system capable of undergoing a Darwinian type of evolution and therefore characterized primarily by its dynamic properties such as metabolism, self-reproduction, and mutability (Miller et al. 1974; Eschenmoser and Kisakürak 1996; Kauffman 1993; Oparin 1965; Eigen 1984). The above criteria, although necessary in defining the phenomenon of life, are clearly not sufficient to establish how inanimate chemistry is transformed into a living system. Based on our present level of comprehension, animate chemistry cannot be defined precisely. In fact, the striving for the complete understanding of living systems, as well as the related challenge of creating an artificial chemical life, fuels our interest and numerous other experimental groups engaged in chemical, biological, and physical sciences. However, despite this incomplete understanding, one can still wonder whether the transition from inanimate to animate chemistry is marked by distinguishing features that may lend themselves to experimental modeling.

Let us consider the living system as an ultimate example of a molecular ecosystem. A molecular ecosystem is defined here as a dynamical self-organized population of interacting and interdependent molecular species that displays emergent properties: a completely novel functional property(s) which was not present beforehand in the individual components of the system. From this perspective, inanimate chemistry belongs to the "molecular world," where the simple sum of the molecular characteristic of each species defines the overall property of a given mixture, while animate chemistry inhabits the realm of the "collective." Therefore, as inanimate chemistry moves toward animate chemistry, a transition must occur from the molecular world into ecosystems which is

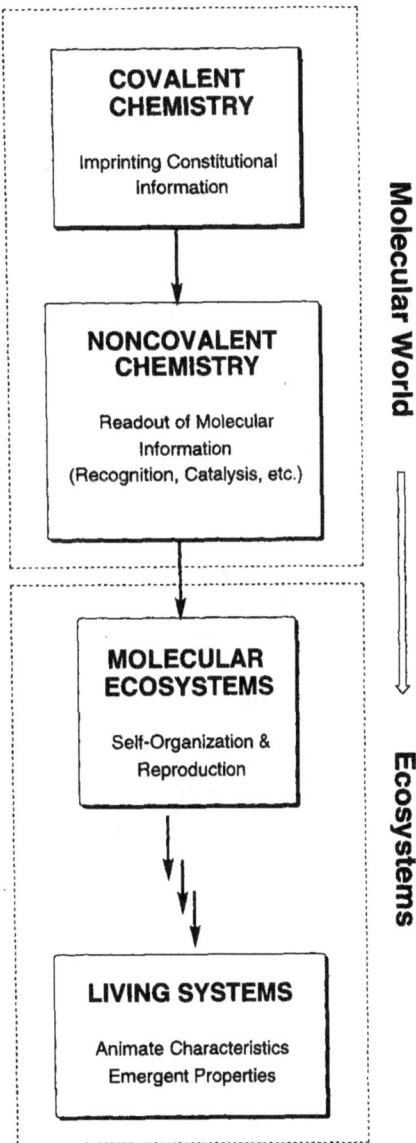

Scheme 1. Hierarchy of information and complexity from inanimate to animate chemistry

signified by the onset of self-replication, self-organization, and the appearance of the phenomenon of emergence. For the purpose of designing suitable chemical model systems, the next critical questions are: what are the necessary requirements for such a transition and what types of molecules can participate in this process? The answers lie in exploring the concept of information and non-linear chemical dynamics in more detail (vide infra).

"Information" is the universal thread that runs through all aspects of chemistry and biology (Küppers 1990). Information originates at the molecular level through covalent chemistry (molecular structure, juxtaposition of functionalities, etc.), is transferred and processed through noncovalent chemistry (recognition, allostery, catalysis, etc.), expands in complexity at the system level (self-organized networks exhibiting emergent properties), and ultimately changes through reproduction and natural selection (Scheme 1). According to Küppers' simplified representation, information can be categorized to have three dimensions in the following order of increasing complexity: syntactic, semantic, and pragmatic (Küppers 1990). Syntactic information denotes the relationship between individual characters or symbols that encode a message. Semantic information, which contains syntactic information, represents what the characters stand for. Finally, pragmatic information, which includes both syntactic and semantic aspects of information, expresses what action the information implies for the sender and recipient. Only molecules that possess pragmatic level information can participate in the formation of self-organized chemical networks. The involvement of pragmatic information in a chemical system can be easily identified "whenever the message or an event, in the widest sense, alters the recipient" (catalysis, allostery, etc.) (Küppers 1990). Therefore, in this respect, substrate-selective catalytic molecular species (enzymes) are prime candidates. However, as mentioned above, a living system is a collective of interacting and interdependent molecular species and thus inherently a "nonlinear" chemical system – as can be easily deduced from its ability to self-reproduce and its various integrated autocatalytic and feedback processes. Therefore, mere linear catalysis is not sufficient, in most instances, to model the process of self-organization. Arguably, the most relevant model systems need to employ informational self-replicating molecular species (Orgel 1992), which provide both the required non-linear chemical catalysis (autocatalytic) and the pragmatic level of molecular information deemed necessary for network self-organization. We suggest that in order to address and ultimately understand the transition process of inanimate to animate chemistry, it is necessary to define basic forms of self-organized autocatalytic chemical networks, how they can be constructed and chemically modeled, and how the interplay of information and nonlinear catalysis can lead to the expression of emergent properties. Here, we briefly review known examples of self-replicating molecular systems, and then focus in more detail on self-organized multicomponent autocatalytic networks.

11.2 Molecular Self-Replication

Self-replication is a special class of autocatalytic reactions in which the product(s) serves as the specific catalyst for its own formation (Lee et al. 1997). Self-replication also implies specificity in the transfer of information to the offspring (product) and thus the requirement for the recognition of substrate(s) during the catalytic transformation. The latter requirement distinguishes self-replication from certain non-informa-

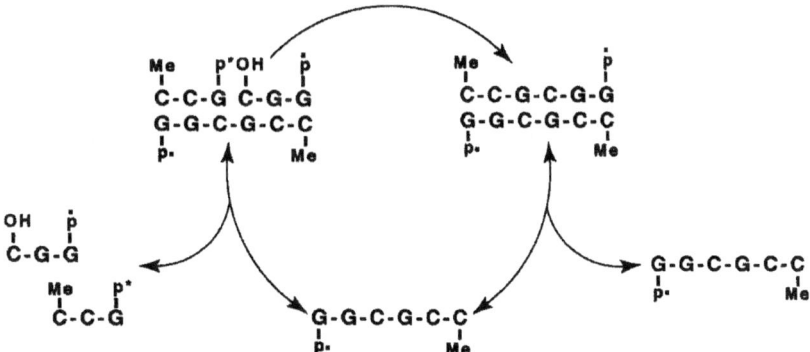

Fig. 1. Self-replicating palindromic hexadeoxynucleotide of von Kiedrowski. The single-stranded template is able to preorganize two constituent fragments, presumably through Watson-Crick base-pairing, to form a ternary complex. In the presence of a water-soluble carbodiimide, this preorganization accelerates the ligation of the two fragments to give an identical copy of the template DNA

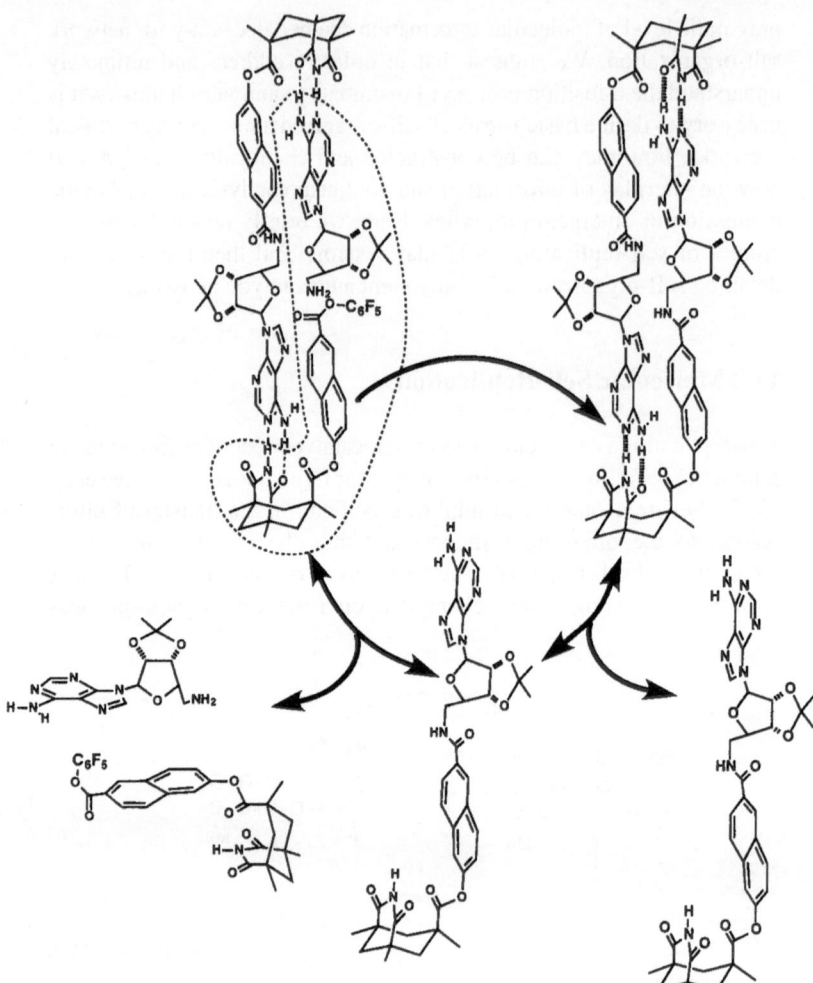

Fig. 2. Self-replication cycle of the abiotic replicator of Rebek. The design of this molecule exploits the ability of the imide derived from Kemp's triacid to selectively bind adenosine through hydrogen-bonding. Joining adenine with this imide via a naphthoyl (or a biphenyl) spacer creates a self-complementary molecule which accelerates its own synthesis from two constituent fragments in a template-dependent manner. The two reactants in the productive ternary complex *are encircled with a dashed line*

tional autocatalytic reactions (Orgel 1992) such as the bromination of acetone (Wintner and Rebek 1996), asymmetric autoinductive reduction of 1,2-diaminoketones (Shibata et al. 1997), and self-reproduction of lipid vesicles (Walde et al. 1994; Morigaki et al. 1997) and micelles (Bachmann et al. 1992). The earliest attempt at non-enzymatic template-directed DNA synthesis dates back to 1966 (Naylor and Gilham 1966). However, it was not until 1986 that von Kiedrowski demonstrated that a palindromic hexanucleotide could catalyze the template-directed ligation of two constituent fragments (Fig. 1) (Kiedrowski 1986). Since then a number of other self-replicating molecules have been designed based on the principles of molecular self-complementarity (Wang and Sutherland 1997; Terfort and Kiedrowski 1992; Pitsch et al. 1995; Perisco and Wuest 1993; Zielinski and Orgel 1987; Bolli et al. 1997; Zhan and Lynn 1997), which include the notable abiotic replicators of Rebek (Fig. 2) (Wintner and Rebek 1996; Tjivikua et al. 1990; Rotello et al. 1991; Park et al. 1992; Nowick et al. 1991; Hong et al. 1992; Feng et al. 1992; Reinhoudt et al. 1996) and, most recently, our own work on self-replicating peptides (Lee et al. 1996; Severin et al. 1997a).

11.3 De Novo Design of Peptide Catalysts

Based on the above perspective, our efforts toward the design and study of self-organized chemical networks necessarily focused on the construction of an informational self-replicating system. From the outset we envisioned that polypeptides could fulfill that role and provide a new class of replicators that could operate in neutral aqueous solutions. The overall concept, which is similar to previously described minimal replicators, required the design of a peptide catalyst for template-directed bimolecular peptide fragment condensation reactions. Our approach began with the design of a peptide ligase (Severin et al. 1997b) – a catalyst for accelerating amide bond formation between two peptide fragments – which we envisioned could be turned into a replicase later on by the appropriate redesign of the catalysts and substrate sequences.

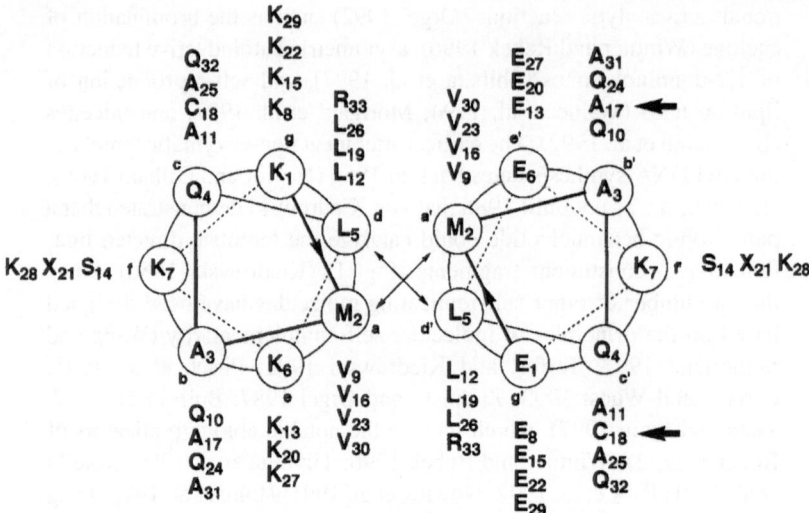

Fig. 3. Helical wheel diagram of the synthetic ligase (*left*) bound to its complementary product (*right*) emphasizing the heptad repeat sequence of coiled-coils. The product was derived from the condensation of two peptide fragments (the reacting residues are Ala-17 and Cys-18, as indicated by *the arrows*). The preponderance of lysine residues at the *e* and *g* positions of the ligase, along with the hydrophobic residues valine and leucine, makes a positively charged molecular surface which is complementary to that of the negatively charged reactants. Furthermore, the positive charges at the molecular recognition surface prevent homodimerization, thereby ensuring that most of the catalyst is available to bind its substrates

11.4 Design of a Synthetic Peptide Ligase

The expression of complementary surfaces required for substrate recognition and catalysis in polypeptides depends not only on the primary sequence but also on the folded three-dimensional structure. Yet, despite remarkable advances in our understanding of protein structure and function, the de novo design of native protein structures remains a formidable challenge (Bryson et al. 1995; Corey and Corey 1996; Corey et al. 1994; Dahiyat and Mayo 1997; Johnson et al. 1993) due to our incomplete understanding of how the primary sequence of polypeptides instruct the folded three-dimensional structures of natural proteins. There-

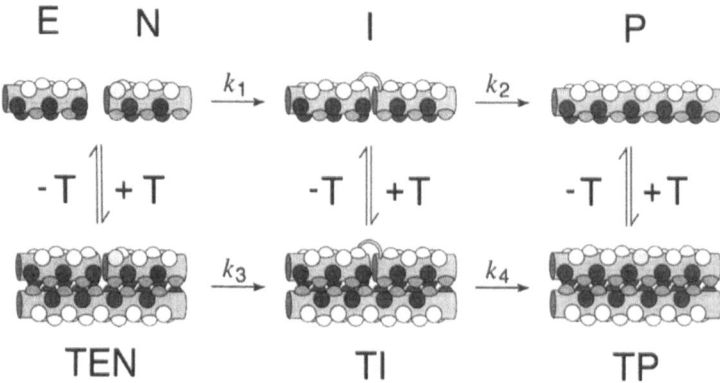

Fig. 4. Minimal reaction model for the ligase catalytic process. The electrophilic and nucleophilic peptide substrates **E** and **N** are preorganized on the complementary template **T**, forming the ternary complex **TEN**. This complex facilitates the condensation reaction (k_3) due to a higher effective concentration of the reactants. The rate-limiting rearrangement (k_4) of the resulting thiolester complex **TI** gives hetero-dimeric coiled-coil **TP**, dissociation of which regenerates the template **T**. The template-independent background reaction is shown above (k_1, k_2). The rearrangement of **TI** is slow in comparison to that of the single-stranded thiolester **I** ($k_4 < k_2$). Peptide backbones are shown as *cylinders* and side chains as *spheres*

fore, from the outset, we chose to base our catalyst designs on the best-understood and the simplest form of protein structure, namely the α-helical coiled-coils (Cohen and Parry 1986, 1990). Coiled-coils are made up of two or more parallel α-helices that wrap one another with a slight left-handed super twist. The primary sequence of coiled-coils exhibits a seven amino acid residue repeat structure (*abcdefg*, Fig. 3). Residues at *a* and *d* (hydrophobic) and *e* and *g* (charged) positions comprise the helical recognition interface, the identity of which has been shown to dictate the thermodynamic and kinetic stability (Wendt et al. 1997; Jelesarov and Bosshard 1996; Hoshino et al. 1997), aggregation state (Harbury et al. 1993, 1994; Gonzalez et al. 1996) and relative orientation of the strands in coiled-coils (Oakley and Kim 1997; Monera et al. 1993, 1994, 1996). We conceptualized that given the cooperative nature of the coiled-coil folding process, one strand may also be effective as a template to preorganize two appropriately preactivated shorter

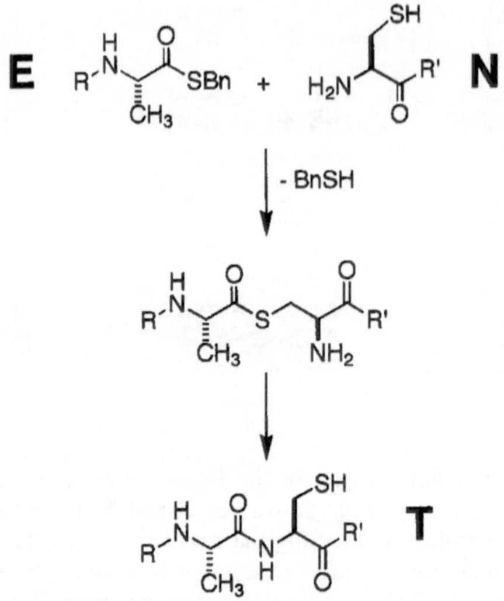

Fig. 5. Formation of a chemo- and regioselective amide bond based on the fragment ligation strategy of Kent. The electrophilic fragment **E** is synthesized as a thiolbenzyl ester and the nucleophilic fragment **N** with an amino-terminal cysteine residue. Fragment condensation initially produces the transthiol esterified intermediate which subsequently rearranges to form a native amide bond

peptide fragments onto itself and thus significantly accelerate their coupling rate (Figs. 4, 5).

This hypothesis was tested recently with a 33-residue peptide sequence (Severin et al. 1997b). Remarkably, the synthetic ligase exhibited rate accelerations of up to 4100 in template-directed ligation of two complementary peptide fragments in neutral aqueous solutions with high sequence and diastereo selectivity and catalytic efficiencies $[(k_{cat}/K_m)/k_{uncat}]$ in excess of 10^5 (Fig. 6). It is also noteworthy that the catalytic efficiency of the synthetic ligase compares favorably with two recently reported catalytic antibodies (Hirschmann et al. 1994; Jacobsen and Schultz 1994; Smithrud et al. 1997) selected for amino acid coupling $[(k_{cat}/K_m)/k_{uncat}$ in the range of 10^2–$10^4]$ and lags by only a factor of 100 behind the reengineered natural enzyme subtilisin (Abrahamsen et al. 1991).

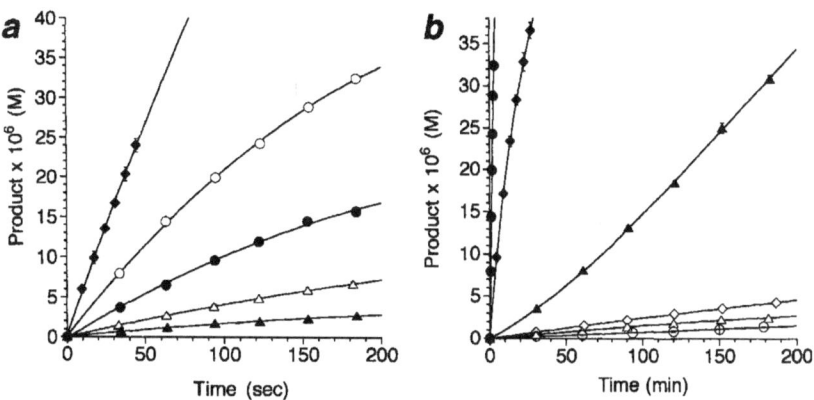

Fig. 6a,b. Experiments demonstrating the catalytic effect of the ligase on peptide fragment condensation. **a** Product formation (**I**) as a function of time for reaction mixtures initially containing 200 µM of the electrophilic substrate, 200 µM nucleophilic substrate, and 11 µM (*filled triangles*), 24 µM (*open triangles*), 53 µM (*filled circles*), 104 µM (*open circles*), and 430 µM (*diamonds*) catalyst **T**, respectively. **b** Product formation (**I+P**) as a function of time for reactions containing equimolar mixtures of various peptide fragments (200 µM each). As the number of complementary charge pairing between residues decreases from 9 (*filled circles*, perfect charge complementarity) to 7 (*diamonds*, 1 charge-charge mismatch and one charge to neutral residue mutation) to 5 (*filled triangles*, 3 charge-charge mismatches and 1 charged to neutral residue mutation), there is a precipitous decrease in the rate acceleration provided by the ligase (50 mol%), from 1800-fold to 82-fold, and 9.2-fold respectively. *Open symbols* represent the formation of the corresponding peptides in the absence of the ligase

11.5 Design of Self-Replicating Peptides

The design of peptide replicases (Lee et al. 1996; Severin et al. 1997a) employs similar principles to those described above but with the following additional consideration. As opposed to the synthetic ligase which was designed to operate based on complementary interactions between the template (catalyst) and its substrates, a replicase must exploit the principles of self-complementarity in order to catalyze formation of a product that is identical in sequence to itself. Fortunately, nature is replete with examples of self-complementary coiled-coil structures (i.e.,

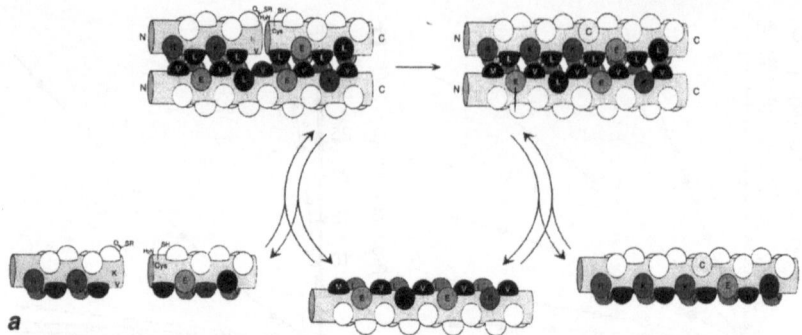

Fig. 7. a Schematic representation of the minimal autocatalytic cycle of a self-replicating peptide. *Cylinders* represent the α-helical peptide backbone while *spheres* represent amino acid side chains. Specific recognition of two constituent fragments by the template leads to the formation of a ternary complex (or quaternary) which promotes peptide fragment condensation, yielding an identical copy of the template. This new copy goes on to establish another autocatalytic cycle. Binding of the fragments is largely mediated by self-complementary interactions between hydrophobic residues (*black and dark gray spheres*) leading to interdigitation of the side chains and a 'knobs-into-holes' type of packing. Crippling the replicator by mutating one of these hydrophobic residues to glutamic acid or even alanine abolishes the self-replication process

homomeric α-helical aggregates). Accordingly, the first self-replicating peptide structure was based on a 32-residue peptide similar in sequence to the leucine-zipper domain of the yeast transcription factor GCN4 (Lee et al. 1996; Severin et al. 1997a; O'Shea et al. 1991). The peptide sequence was modified to allow the possibility of autocatalysis through one- and/or two-stranded α-helical template structures. Peptide substrates, the 17-residue electrophilic and a 15-residue nucleophilic fragments, were in turn designed to be the constituent fragments of the template sequence with the ligation site placed on the solvent-exposed surface of the coiled-coil structure in order to avoid interference with the helical recognition interface.

A priori, a certain dynamic characteristic is expected for minimal self-replicating molecules (Kiedrowski 1993). In the absence of any template (product) in the initial reaction mixture, product formation ensues through bimolecular (background) fragment condensations. The

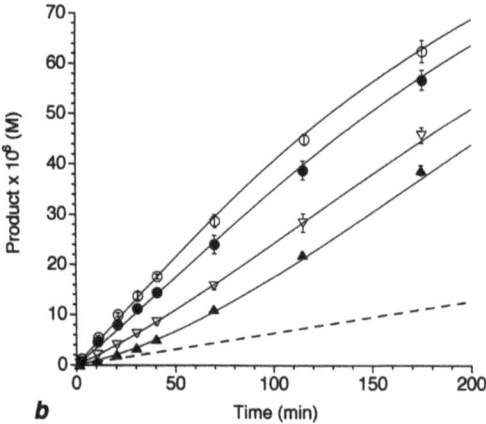

Fig.7. b Template formation as a function of time for reaction mixtures initially containing 0% (*filled triangles*), 5% (*open triangles*), 15% (*filled circles*), and 30% (*open circles*) template, respectively. Curves were generated by non-linear least squares fitting of the data to the empirical equation of von Kiedrowski. *The dashed line* represents the bimolecular production of template in the absence of autocatalysis, which agrees well with rates observed for reactions carried out under denaturing conditions. *Error bars* reflect standard deviations of three independent runs

reaction which would be relatively slow in the beginning should increase in rate as more product (catalyst) is produced in the reaction mixture. In an efficient self-replication process, this behavior typically manifests itself in a sigmoidal growth profile. In practice autocatalysis can be easily identified by the fact that if the product of the reaction is truly a catalyst for the reaction, then addition of the product to the reaction mixture should lead to greater initial rates of product formation. Similarly in our system, kinetic analyses of reaction mixtures differing only in the initial concentration of template were used to unequivocally establish the autocatalytic nature of the process (Fig. 7). The requirement for peptide folding and molecular recognition in the template-directed ligation process was revealed by the absence of autocatalysis under denaturing solution conditions (Severin et al. 1997a). Moreover, the intermediary of complexes in which both substrates are recognized by the template (the "true" self-replicating complexes) was established

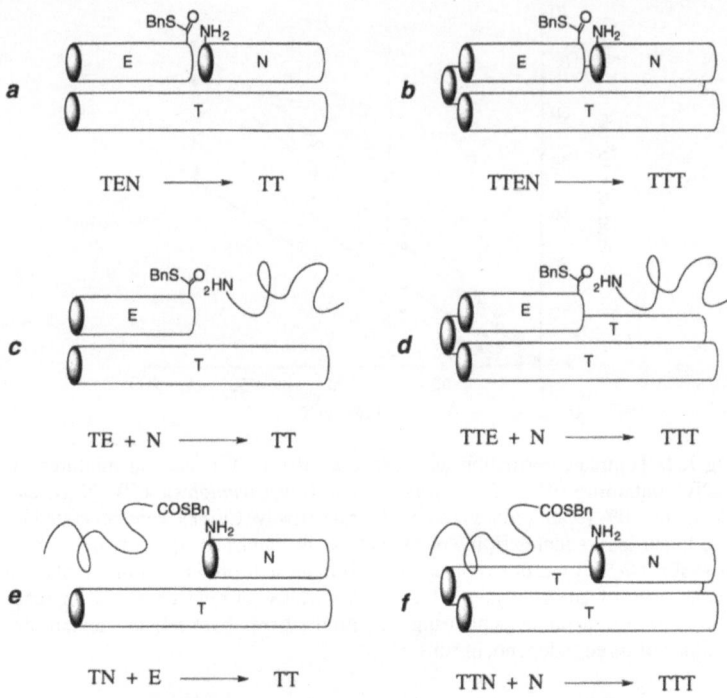

Fig. 8a–f. Schematic representation of the plausible intermediates which can contribute to the production of the replicator **T**. Ligation via the ternary or quaternary complexes **TEN** or **TTEN** (**a,b**) does involve molecular recognition of both fragments and can therefore be considered "true" self-replication intermediates (Reinhoudt et al. 1996). For the bimolecular reactions shown in **c–f** only one peptide fragment is structurally organized on the template. Therefore these reactions are 'merely' autocatalytic. Control experiments with crippled templates indicate that intermediates **c–f** are not involved in the autocatalytic process

through control experiments with crippled templates (Fig. 8). Templates in which the binding of either the electrophilic or nucleophilic substrate was disabled by single glutamic acid substitution within the corresponding hydrophobic recognition interface, displayed no significant catalytic contribution to the initial rate of ligation. Interestingly, the peptide self-replication also exhibited high sequence fidelity. Even conservative single amino acid mutations – alanine in place of leucine or valine – in the hydrophobic recognition interface lead to abolition of the autocatalytic process.

In molecular self-replication, the dependence of the initial rates of product formation on the initial template concentration is of special significance (Kiedrowski 1993). In the absence of product inhibition, self-replicating molecular species should display exponential growth profiles. In other words, the initial rate of product formation should scale linearly with the initial concentration of the replicator present in the reaction mixture. However, because during the catalytic cycle the product-template complex produced is often more stable than the complex from which it was derived (ca. the ternary complex), the reaction becomes product inhibited, which significantly lowers the concentration of available catalyst for the next catalytic cycle. In practice, instead of the exponential growth characteristics, minimal replicators often exhibit parabolic growth profiles. This pattern is characterized by the linear dependence of the initial rate of product formation on the square root of the initial template concentration (square-root law) and can be modeled by the empirical rate equation of Kiedrowski (1986): $d(T)/dt=(N_0-T)(E_0-T)[k_a (T+T_0)^p+k_b]$. According to this equation, the initial rate of template formation (T) is described in three terms: a template-independent term (background) with the apparent rate constant of k_b, a template-dependent term (autocatalytic) with a rate constant k_a and the reaction order p (N_0, E_0, and T_0 are the initial concentrations of the nucleophile, electrophile, and the template, respectively). The reaction order p=0.5 describes the typical behavior of parabolic replicators and p=1 the limiting exponential growth characteristics. Interestingly, depending on the initial concentration of the peptide fragments employed, the peptide replicator displayed significant variations in its growth profile in terms of both efficiency ($\epsilon=k_a/k_b$) and the reaction order p. At 9×10^{-5} M electrophile and nucleophile concentration, the reaction growth profile obeyed the expected square-root law (p=0.5, $\epsilon=470$ M$^{-0.5}$) (Lee et al. 1996). But at a twofold higher concentration (18×10^{-5} M), a higher order growth pattern and efficiency were observed (p=0.63, $\epsilon=3700$ M$^{-0.6}$) (Severin et al. 1997a). Kinetic modeling suggests that higher order aggregates (quaternary complexes; Fig. 3b) in which the template dimer is the active catalytic species may also contribute to the overall autocatalytic process (Severin et al. 1997a). Chmielewski and coworkers have recently described a pH-modulated

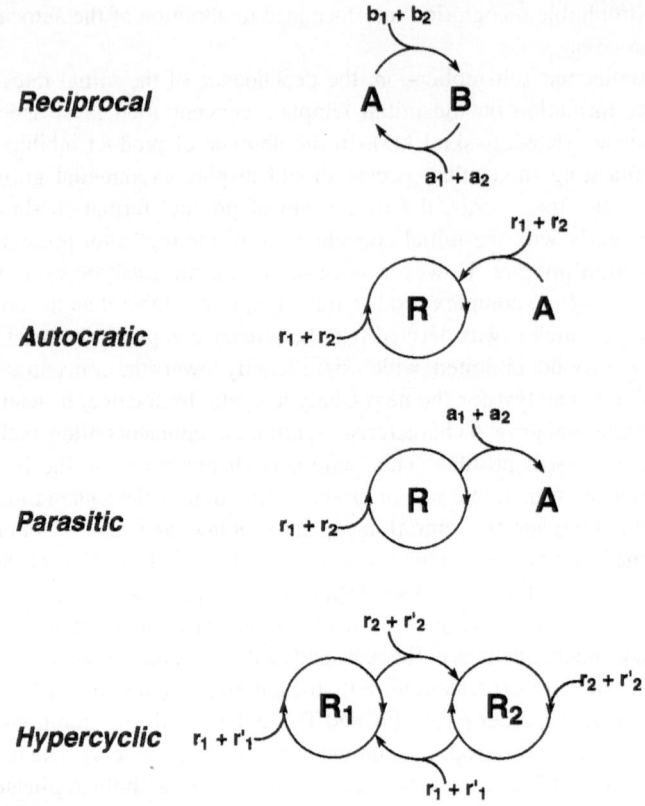

Scheme 2. Primary forms of self-organized autocatalytic networks

variant of the self-replicating peptide (Yao et al. 1997). The peptide sequence was designed with glutamic acid residues at all e and g positions of the heptad repeat in order to disfavor helical assembly in neutral to basic reaction conditions. The replicator was thus shown to be autocatalytically inactive at pH$\geq$5 but displaying efficient catalysis at pH=4 (p=0.5, ε=900 M$^{-0.5}$). The design of self-replicating peptides that can function under different solution conditions bodes well for the versatility of coiled-coils in this regard.

It is worth pointing out that from the perspective of the origin-of-life, the dynamics of replicator growth are critically important. As pointed out by Szathmary, the population of competing parabolic replicators leads to "survival of everyone," i.e. coexistence rather than selection (Szathmary and Gladkih 1993). Nevertheless, for the present purpose of designing nonlinear chemical networks, even though no species can be driven to extinction, the relative efficiency of competing replicators and the initial starting reaction conditions should play a significant role in determining which species would dominate in the reaction mixture.

11.6 Self-Organized Autocatalytic Networks

At the primary level, one can imagine at least four basic forms of self-organized autocatalytic networks: *reciprocal, autocratic, parasitic, and hypercyclic* (Scheme 2). Each network is endowed with unique interaction dynamics that can manifest in the emergence of various collective behaviors.

11.7 Reciprocal Networks

Self-reproduction at the system level is not restricted to self-comple-mentary structures (Kauffman 1993). In a multicomponent reaction mixture, if the formation of each molecular species is catalyzed by another member of the population, then reproduction becomes an emer-gent property of the system as a whole. The minimal form of a recipro-cal network is made up of two members with each templating, through complementary interactions, the formation of the other (Scheme 2). Nucleic acid structures are obvious candidates for such studies since one strand has the complete information to dictate the formation of the complementary strand and vice versa. Indeed, replication through cross catalysis by two short oligonucleotides has been studied (Sievers and Kiedrowski 1994). Sievers and von Keidrowski studied a system based on four trideoxynuleotide derivatives (Fig. 9). The fragments undergo competitive condensation reactions in the presence of water-soluble carbodiimide to produce two self-complementary hexanucleotides as well as two complementary hexanucleotides. Careful studies suggested that the formation of the two complementary hexanucleotides is acceler-ated due to cross-catalysis. However, the two self-complementary se-

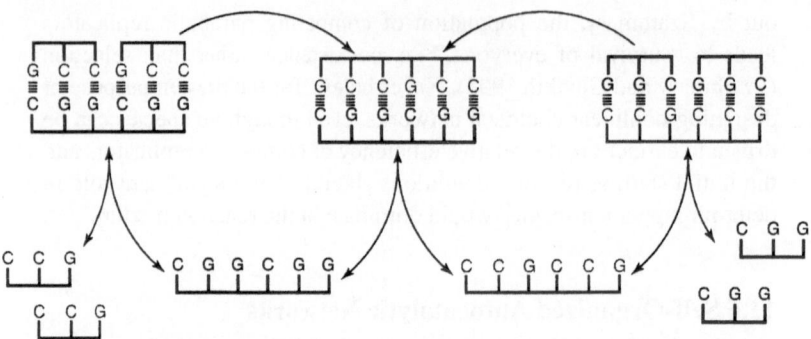

Fig. 9. Cross-catalytic cycle of two complementary DNAs. A hexadeoxynucleotide binds two complementary fragments thereby promoting their fragment condensation. This new strand, in turn, can catalyze the ligation of its own complementary substrates to produce the original hexadeoxynucleotide template. Hence, reproduction arises as the result of reciprocal catalysis mediated by complementary sequences. The fragments also combine in two other ways to form self-complementary sequences which are capable of self-replication but not cross-catalysis

quences were also shown to catalyze their own formation with efficiencies similar to that of the cross-catalytic process.

Reciprocal catalysis has also been investigated in other molecular systems. Rebek and coworkers have established, in independent reactions, the cross-catalytic formation of complementary molecular structures (Pieters and Huc 1994, 1995). Eschenmoser and coworkers recently demonstrated that a variety of pRNA sequences are capable of template-directed catalysis of complementary structures (Pitsch et al. 1995; Bolli et al. 1997). Likewise, Bohler et al. (1995) have shown that poly-C-containing PNA (peptide nucleic acid) or DNA can be used to catalyze the production of poly-G RNA or PNA, respectively. Coiled-coil peptides have also been used to cross-catalyze formation of complementary sequences (Y. Yokobayashi, D.H. Lee, M.R. Ghadiri, unpublished). However, in most instances, either due to shunt pathways or severe product inhibition, reciprocal catalysis has not been realized in one pot. As will be described, our group has demonstrated the feasibility of reciprocal peptide catalysis in the context of a putative hypercyclic network (Lee et al. 1997).

11.8 Autocratic and Parasitic Networks

An autocratic network is established when a self-replicating cycle(s) is enhanced unilaterally by the specific catalytic action of another species. The reverse situation defines the parasitic network. An autocratic network represents a positively regulated form of molecular self-replication, while a parasitic network may be thought of as the corresponding negative regulatory form (Scheme 2). An autocratic network was discovered in a reaction mixture composed of four peptide fragments: the electrophilic **E** and nucleophilic **N** fragments of a replicator and their corresponding single alanine mutants **E$_{9A}$** and **N$_{26A}$**, respectively (Severin et al. 1998). The peptide fragments underwent competitive

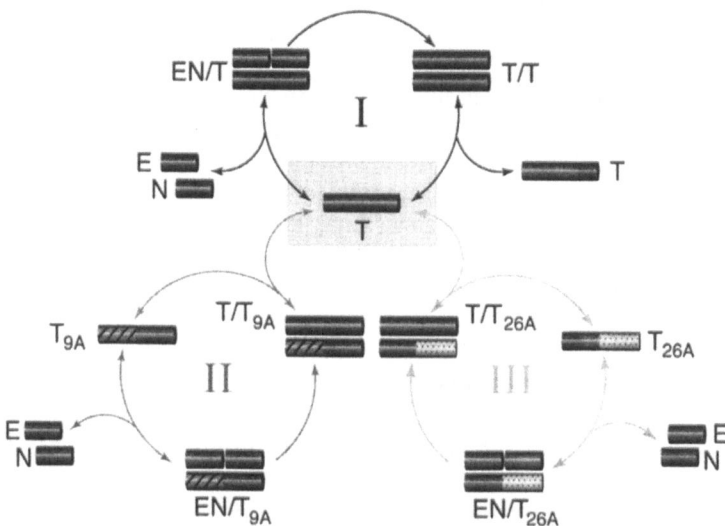

Fig. 10. Mixing four peptide fragments: electrophile **E**, nucleophile **N**, along with their respective mutants **E$_{9A}$** and **N$_{26A}$**, results in the formation of a self-organized network composed of three interconnected (auto)catalytic cycles that displays overall dynamic error-correcting properties. The mutant products **T$_{9A}$** (*striped-solid*) and **T$_{26A}$** (*solid-speckled*) do not self-replicate but do catalyze the formation of the native replicator **T** (*solid*). This selective upregulation results in enhanced reproduction of T at the expense of the mutants, which are made from common fragments, thereby providing a certain degree of stabilization to the wild-type "peptide genome"

ligation reactions to produce predominantly the native replicator T with the three mutant peptides T_{9A}, T_{26A}, and $T_{9/26A}$. Several competition and control experiments revealed that the native replicator became the dominant species in the reaction mixture in part because the mutants were autocatalytically infertile. However, the single mutants T_{9A} and T_{26A} were both surprisingly good catalysts for the production of the native replicator sequence. Control experiments indicated that these mutants catalyzed formation of T with efficiencies approximately 75% of the autocatalytic channel as judged by the initial rates of template formation in the presence of similar amounts of added catalysts. Interestingly, the replicator did not catalyze formation of either mutant. Therefore, the simple four-component reaction mixture self-organized to give a network made up of three interconnected (auto)catalytic cycles (Fig. 10), through which the native replicator, in a sense, subjugated the mutants for its own production. The net effect is a significant upregulation in the production of the native replicator in response to the appearance of mutants (errors) in the reaction mixture. This particular autocratic process, which has been dubbed "dynamic error correction" (Severin et al. 1998), may have significant implications for genotype stabilization in certain settings (note that for an absolute error-correcting process, not just in the dynamic sense meant here, both instructed bond-making and bond-breaking events are required for sequence editing). To date, no example of a parasitic network has been reported.

11.9 Hypercyclic Networks

The class of hypercyclic organization was conceived originally by Manfred Eigen in 1971 (Eigen and Schuster 1979) and then further developed by Eigen and Schuster (1977, 1978a,b, 1979). Their seminal studies launched the "molecular-Darwinistic" approach toward the origin of life issues. It postulates that "biological information is the result of a spontaneous process in the course of which inanimate matter organizes itself of its own accord into animate systems, by learning from the conditions of its surroundings and developing ever higher degrees of complexity and organization" (Küppers 1990). Clearly, many of the perspectives presented in this account are influenced by this principle of self-organization. At the heart of Eigen's postulate lies the hypercyclic

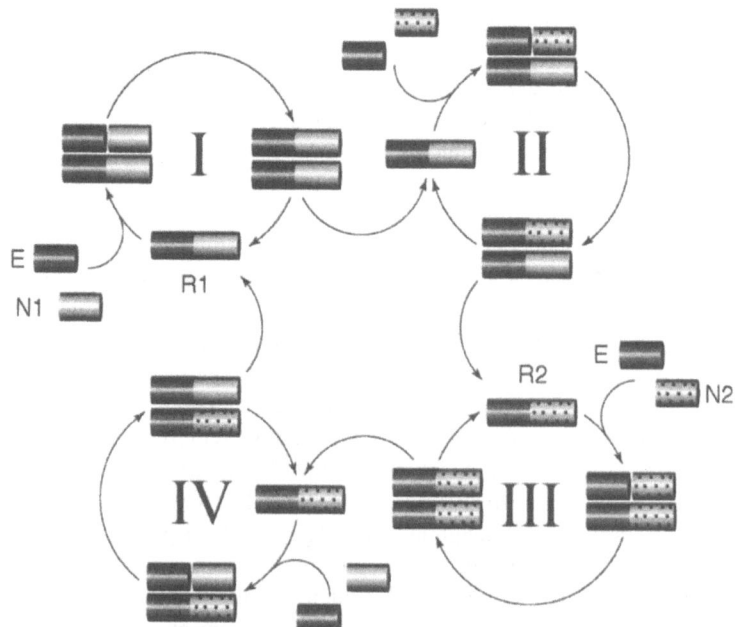

Fig. 11. Self-replicating peptide **R₁** (*dark/light gray*, autocatalytic cycle I), made from the ligation of **E** (*dark gray*) with **N₁** (*light gray*), and replicator **R₂** (*dark gray/speckled*, autocatalytic cycle III), derived from the **E** and **N₂** (*speckled*), link through the cross-catalytic cycles II and IV to establish a mutualistic network. This form of self-organization is stabilized by virtue of the fact that each replicator makes the other member better than they reproduce themselves. The mechanistic details of this network may be more complex than the minimal system depicted here and may involve the intermediary of higher order complexes in the catalytic and autocatalytic processes. For instance, the involvement of both **R₁** and **R₂** in a quaternary complex in cycles II and IV would establish the second-order autocatalysis required by a hypercyclic network

form of self-organization. Hypercycles are defined as being present when two or more replicators are connected by another superimposed autocatalytic cycle. Therefore, a hypercycle is based on second-order (or higher) autocatalysis which manifests itself in the hyperbolic (faster than exponential) growth characteristics of its members. It constitutes the essence of mutualism and symbiosis at the molecular level. Recently, we reported a self-organized system that displayed some of the characteristics of hypercycles (Lee et al. 1997). The system is composed of two replicators each catalyzing the formation of the other; thus the replicators are linked by a superimposed catalytic cycle (Fig. 11). Interestingly, this relatively complex form of mutualistic self-organization is brought about simply by the reactions of two nucleophilic fragments N_1 and N_2 with one electrophilic fragment E. Fragment condensation reactions in neutral aqueous solutions yield two replicators R_1 and R_2 ($\varepsilon_{R1}=470$ $M^{-0.5}$, $\varepsilon_{R2}=720$ $M^{-0.5}$). A priori, a survival-of-the-fittest may be expected where the more efficient replicator R_2 would overwhelm R_1 by its more rapid consumption of the common fragment E. However, the situation is both more interesting and complex. When the initial reaction mixtures were seeded with either R_1 or R_2, surprisingly the production of both replicators was enhanced. More interestingly, R_1 cross-catalyzed the production of R_2 more efficiently that its own formation. The reverse was also true: R_2 was found to be a more efficient catalyst for R_1 than for itself. Therefore, the system although not a true hypercycle displays mutualism among otherwise competitive self-replicating species.

In summary, the experimental evidence presented here seems to support the hypothesis that informational nonlinear chemical systems are sufficient for modeling the process of self-organization and the initial steps of the transition from simple molecular mixtures into molecular ecosystems. However, even a cursory examination of what is known about the chemistry of living systems makes it clear that higher levels of self-organization are necessary for the transition of inanimate to animate matter. Clearly, at the present stage of our understanding and the availability of only simplistic model systems, several other basic questions need to be addressed. What other primary or secondary forms of self-organized networks are possible? How can several subsystems be productively integrated? What emergent properties might these systems display? Can nucleic acids and polypeptides participate cooperatively in network self-organization? The challenges are to design and characterize new and more complex non-linear chemical networks and in time even molecular ecosystems which may help us to better understand the complex web of living chemistry. It is our hope that the working perspectives outlined here will serve to catalyze further discussions and lead to more advanced experimental and theoretical research in this field.

Acknowledgements. I would like to thank my colleagues Professor Julius Rebek and Albert Eschenmoser for stimulating discussions related to the topics described here. I also thank our colleagues M. Churchill, J.R. Granja, A.J. Kennan, K. Kumar, D.H. Lee, J.A. Martinez, K. Severin, and Y. Yokobayashi for their invaluable contributions to the research program described.

References

Abrahamsen L, Tom J, Burnier J, Butcher KA, Kossiakoff A, Wells JA (1991) Engineering subtilisin and its substrates for efficient ligation of peptide bonds in aqueous solution. Biochemistry 30:4151–4159

Bachmann PA, Luisi PL, Lang J (1992) Autocatalytic self-replicating micelles as models for prebiotic structures. Nature 357:57–59

Bohler C, Nielsen PE, Orgel LE (1995) Nature 376:578–581

Bolli M, Micura R, Pitsch S, Eschenmoser A (1997) Further observations on replication. Part 5. Helv Chim Acta 80:1901–1951

Bryson JW, Betz SF, Lu HS, Suich DJ, Zhou HX, O'Neil K, DeGrado WF (1995) Protein design: a hierarchic approach. Science 270:935–941

Cohen C, Parry DAD (1986) Helical coiled coils – a widespread motif in proteins. Trends Biochem 11:245–248

Cohen C, Parry DAD (1990) Proteins: structure, function, and engineering 7:1–15

Corey MJ, Corey E (1996) On the failure of de novo-designed peptides as biocatalysts. Proc Natl Acad Sci USA 93:11428–11434

Corey MJ, Hallakova E, Pugh K, Stewart JM (1994) Studies on chymotrypsin-like catalysis by synthetic peptides. Appl Biochem Biotechnol 47:199–210

Dahiyat BI, Mayo SL (1997) De novo protein design: fully automated sequence selection. Science 278:82–87

Eigen M (1984) The origin and evolution of life at the molecular level. Adv Chem Phys 55:119–137

Eigen M, Schuster P (1977) The hypercycle. A principle of natural self-organization. Part A: emergence of the hypercycle. Naturwissenschaften 64:541–565

Eigen M, Schuster P (1978a) The hypercycle. A principle of natural self-organization. Part B. Naturwissenschaften 65:7–41

Eigen M, Schuster P (1978b) The hypercycle. A principle of natural self-organization. Part C: the realistic hypercycle. Naturwissenschaften 65:341–369

Eigen M, Schuster P (1979) The hyercycle. A principle of natural self-organization. Springer, Berlin

Eschenmoser A, Kisakürek MV (1996) Helv Chim Acta 79:1249–1259

Feng Q, Park TK, Julius Rebek J (1992) Crossover reactions between synthetic replicators yield active and inactive recombinants. Science 256:1179–1180

Gonzalez L, Brown RA, Richardson D. Alber T (1996) Buried polar residues and structural specificity in the GCN4 leucine zipper. Nature Struct Biol 3:1002–1010

Harbury PB, Zhang T, Kim PS, Alber T (1993) A switch between two-, three-, and four-stranded coiled coils in GCN4 leucine zipper mutants. Science 262:1401–1407

Harbury PB, Kim PS, Alber T (1994) Crystal structure of an isoleucine-zipper trimer. Nature 371:80–83

Hirschmann R III, Smith AB III, Taylor CM, Benkovic PA, Taylor SD, Yager KM, Sprengler PA, Benkovic SJ (1994) Peptide synthesis catalyzed by an antibody containing a binding site for variable amino acids. Science 256:234–237

Hong J-I, Feng Q, Rotello V, Julius Rebek J (1992) Competition, cooperation, and mutation: improving a synthetic replicator by light irradiation. Science 255:848–850

Hoshino M. Yumoto N, Yoshikawa S, Goto Y (1997) Design and characterization of the anion-sensitive coiled-coil peptide. Protein Sci 6:1396–1404

Jacobsen JR, Schultz PG (1994) Antibody catalysis of peptide bond formation. Proc Natl Acad Sci USA 91:5888–5892

Jelesarov I, Bosshard HR (1996) Thermodynamic characterization of the coupled folding and association of heterodimeric coiled coils (leucine zippers). J Mol Biol 263:344–358

Johnsson K, Allemann RK, Widmer H, Benner SA (1993) Synthesis, structure and activity of artificial, rationally designed catalytic polypeptides. Nature 365:530–532

Kauffman SA (1993) The origins of order. Oxford University Press, New York

Kiedrowski G von (1986) A self-replicating hexadeoxynucleotide. Angew Chemie (International Edition in English) 25:932–935

Kiedrowski G von (1993) Minimal replicator theory. I. Parabolic versus exponential growth. Bioorganic Chem Frontiers 3:113–146

Küppers B-O (1990) Information and the origin of life. MIT Press, Cambridge, MA

Lee DH, Granja J, Martinez JA, Severin K, Ghadiri MR (1996) A self-replicating peptide. Nature 382:525–528

Lee DH, Severin K, Ghadiri MR (1997) Autocatalytic networks: the transition from molecular self-replication to molecular ecosystems. Curr Opin Chem Biol 1:491–496

Lee DH, Severin K, Yokobayashi Y, Ghadiri MR (1997) Emergence of symbiosis in peptide self-replication through a hypercyclic network. Nature 390:591–594

Li T, Nicolaou KC (1994) Chemical self-replication of palindromic duplex DNA. Nature 369:218–221

Miller SL, Orgel LE (1974) The origins of life on earth. Prentice Hall, Englewoods Cliff, NJ

Monera OD, Zhou NE, Kay CM, Hodges RS (1993) Comparison of antiparallel and parallel two-stranded α-helical coiled-coils. Design, synthesis, and characterization. J Biol Chem 268:19218–19227

Monera OD, Kay CM, Hodges RS (1994) Electrostatic interactions control the parallel and antiparallel orientation of α-helical chains in two-stranded α-helical coiled-coils. Biochemistry 33:3862–3871

Monera OD, Zhou NE, Lavigne P, Kay CM, Hodges RS (1996) Formation of parallel and antiparallel coiled-coils controlled by the relative positions of alanine residues in the hydrophobic core. J Biol Chem 271:3995–4001

Morigaki K, Dallavalle S, Walde P, Colonna S, Luisi PL (1997) Autopoietic self-reproduction of chiral fatty acid vesicles. J Am Chem Soc 119:292–301

Naylor R, Gilham PT (1966) Biochemistry 5:2722–2728

Nowick JS, Feng Q, Tjivikua T, Ballester P, Julius Rebek J (1991) Kinetic studies and modeling of a self-replicating system. J Am Chem Soc 113:8831–8839

Oakley MG, Kim PS (1997) Protein dissection of the antiparallel coiled coil from Escherichia coli seryl tRNA synthetase. Biochemistry 36:2544–2549

Oparin AI (1965) Adv Enzymol 27:347–380

Orgel LE (1992) Molecular replication. Nature 358:203–209

O'Shea EK, Klemm JD, Kim PS, Alber T (1991) X-ray structure of the GCN4 leucine zipper, a two-stranded, parallel coiled coil. Science 254:539–544

Park TK, Feng Q, Rebek J (1992) Synthetic replicators and extrabiotic chemistry. J Am Chem Soc 114:4529–4532

Perisco F, Wuest JD (1993) Use of hydrogen bonds to control molecular aggregation. Behavior of a self-complementary dipyridone designed to self-replicate. J Org Chem 58:95–99

Pieters RJ, Huc I, Rebek J Jr (1994) Reciprocal template effects in a replication cycle. Angewandte Chemie (International Edition in English) 33:1579–1581

Pieters RJ, Huc I, Rebek J Jr (1995) Reciprocal template effects in bisubstrate systems: a replication cycle. Tetrahedron 51:485–498

Pitsch S, Krishnamurthy R, Bolli M, Wendeborn S, Holzner A, Minton M, Lesueur C, Schlonvogt I, Jaun B, Eschenmoser A (1995) Pyranosyl-RNA ('p-RNA'): base-pairing selectivity and potential to replicate. Helv Chim Acta 78:1621–1635

Reinhoudt DN, Rudkevich DM, Jong FD (1996) Kinetic analysis of the Rebek self-replicating system: is there a controversy?J Am Chem Soc 118:6880–6889

Rotello V, Hong J-I, Julius Rebek J (1991) Sigmoidal growth in a self-replicating system. J Am Chem Soc 113:9422–9423

Severin K, Lee DH, Martinez JA, Ghadiri MR (1997a) Peptide self-replication via template-directed ligation. Chem Eur J 3:1017–1024

Severin K, Lee DH, Kennan AJ, Ghadiri MR (1997b) A synthetic peptide ligase. Nature 389:706–709

Severin K, Lee DH, Vieth M, Ghadiri MR (1998) Dynamic error correction in autocatalytic peptide networks. Angew Chemie (International Edition in English) 37:126–128

Shibata T, Takahashi T, Konishi T, Soai K (1997) Asymmetric self-replication of chiral 1,2-amino alcohols by highly enantioselective autoinductive reduction. Angew Chemie (International Edition in English) 36:2458–2460

Sievers D, Kiedrowski G von (1994) Self-replication of hexadeoxynucleotide analogs: autocatalysis versus cross-catalysis. Nature 369:221–224

Smithrud DB, Benkovic PA, Benkovic SJ, Taylor CM, Yager KM, Witherington J, Philips BW. Sprengler PA III, ABS, Hirschmann R (1997) Investigations of an antibody ligase. J Am Chem Soc 117:278–282

Szathmary E, Gladkih I (1993) Subexponential growth and coexistence of nonenzymically replicating templates. J Theoret Biol 138:55–58

Terfort A, Kiedrowski G (1992) Self-replication by condensation of 3-aminobenzamidines with 2-formylphenoxy acetic acids. Angew Chemie (International Edition in English) 31:654–656

Tjivikua T, Ballester P, J Rebek J (1990) Self-replicating system .J Am Chem Soc 112:1249–1250

Walde P, Wick R, Fresta M, Mangone A, Luisi PL (1994) Autopoietic self-reproduction of fatty acid vesicles. J Am Chem Soc 116:11649–11654

Wang B, Sutherland IO (1997) Self-replication in a Diels-Alder reaction. Chemical Communications 1495–1496

Wendt H, Leder L, Haermae H, Jelesarov I, Baici A. Bosshard HR (1997) Very rapid, ionic strength-dependent association and folding of a heterodimeric leucine zipper. Biochemistry 36:204–213

Wintner EA, Rebek J (1996) Autocatalysis and the generation of self-replicating systems. Acta Chim Scand 50:469–485

Yao S, Ghosh I, Zutshi R, Chmielewski J (1997) A pH-modulated, self-replicating peptide. J Am Chem Soc 119:10559–10560

Zhan Z-YJ, Lynn DG (1997) Chemical Amplification through template-directed synthesis. J Am Chem Soc 119:12420–12421

Zielinski WS. Orgel LE (1987) Autocatalytic synthesis of a tetranucleotide analog. Nature 327:346–347

Subject Index

1,1'-binaphthalene   64
1,1'-binaphthalene-2,2'-diol   58, 65
1,1'-binaphthalene-based receptors
    58
1-octyl β-d-glucoside   65
1-octyl pyranosides   65
^{15}N labels   36
^{15}N-enriched ε nitrogen   45
^{19}F-labeled tryptophan   36
^{1}H-NMR titration binding studies
    69
3$_1$ helix   46
^{31}P-containing ligands   36, 40
5-enolpyruvylshikimate-3-phos-
    phate   35
6-(p-toluidino)naphthalene-2-sulfon-
    ate   72
9,9'-spirobi   55

α helix   46
ab initio calculations   114
abiological catalysis   23
acceptors   113
alkylation reactions   2
allosteric effectors   98
allylic alkylations   2
amide groups   100
anti conformation   122
antigene   120, 123, 127
antigene therapy   164
antisense   119, 127

aptamers   120, 135
arabinose binding protein   63
arenes   72
aromatic interactions   62
asymmetric amino acid synthesis
    20
asymmetric catalysis   2
asymmetric desymmetrization   7
atherosclerotic plaque   76
autocalytic   214, 217, 225
automatic docking   106

β-estradiol   71
β-strand   46
bidentate H-bonding   55
bifurcated H-bond   57
bile acids   71
binding and selection of substrate
    97
binding motifs   99
bioassays for enzymes   173
biochemical pathways   98
biomedicinal technologies   76
biomolecular recognition   206
bis-PNA   159
bridge motion   46

cage-type receptors   54
Cambridge Crystallographic Data-
    base   109

Cambridge Structural Database
    102
carbohydrates   184
carbohydrate binding proteins   54,
    169
carbohydrate libraries   170
carbohydrate recognition   63
carbohydrate recognition do-
    main   170
catalytic dendrophanes   76
catalytic nucleic acids   143
catalytic RNA   135
cell adhesion   206
cell wall   48
cell-wall attachment   48
chelate effect   189, 194, 196, 198,
    205
chiral drugs   58
chiral recognition   60
chiral space   11
chiral stationary phase   57
cholera toxin   172
cholesterol deposits   76
cholesterol dissolution   69
circular oligonucleotides   120, 129
cis-carbamate conformation   61
clamp   120, 126
cleft   35
cleft-type receptors   54, 55
collectins   170
combinatorial selection   135
complexation of sugars   63
computer simulation   61
conformation   44
conformation space   112
conformational analysis   115
conformational flexibility   107, 112
conformational homogeneity   63
conformational preferences   102,
    108, 109
conformational properties   99

conformationally constrained
    ligands   10
conical receptor   71
constraints   33
cortisone   71
cross-coupling   66
cross-linking   43
crystal packing   113
crystal-field environment   103
cyclodextrins   71
cyclophane   71
cyclophane core   76
cyclophane receptors   54, 64, 71

D-Ala-D-Ala terminus   47
de novo design   99, 115
decoy   119
dendritic receptors   72
dendritic superstructure   72
dendrophanes   72
dephasing   27
diaminopurine   163
difference spectroscopy   61
dihedral angle θ   61
dipolar coupling   26
dispersion interactions   66
distamycin   152
DNA   151
DNA targeting   154
docking   99, 113, 115
double duplex invasion   163
double-decker cyclophane recep-
    tors   66
drug-DNA adduct   83, 89

E-selectin   171
E. coli expression system   36
Ecteinascidins   81
electrostatic O···H interactions   61
elongation factor Tu (EF-Tu)   29
enantioselective recognition   55
enzymes   98

ester transferase ribozyme 144
ethambutol 12
excitatory amino acid 54
excitatory amino acid derivatives 58

face-to-face stacking 62
FlexX 106, 113
fluorescence binding titrations 72
fluorescence relaxation measure-
    ments 75
freeze-quenched buffer 36
functional group desolvation 71
functional-group replacements
    113, 114

G-quartet 138
global minimum 108
globular dendrimers 54
globular proteins 54, 72
glutamine binding protein (GlnBP)
    33
– cleft 34
glycolipids 184
glycopeptide libraries 172
glycopeptide template libraries 173
glycopeptide templates 171
glycoproteins 184
glycosidic bond 66
glycosyl amino acid building
    blocks 172
glycosylated Fmoc-amino acid-
    OPfp esters 174
Gulliver Effect 47

H-bonding motifs 61
hepatic lectin 177
hetereonuclear dipolar coupling 28
Hiyama coupling 66
host-guest association 55
host-guest exchange kinetics 65,
    72, 76
hydrated whole cells 45

hydrocortisone 71
hydrogen-bonding ability 113
hydrogen-bonding properties 113
hydrophobic desolvation 63, 66
hypothetical receptor models 103

in situ capping 173
in vitro selection 135
in vitro transcription 135
inclusion complexes 71
intermolecular interactions 97
ionic H-bonding 65
ionic H-bonding residues 63
isoprenoidal side chain 71
IsoStar 106, 114
isosteric functional group replace-
    ment 115

Job plot analysis 66

L-arabinose 63
ladder sequencing 173
ladder-encoded glycopeptide li-
    brary 175
Lathyrus odoratus 175
lectins 194, 196, 200, 205
LUDI 106

MacroModel 55, 61
magic angle 26
magic-angle sample spinning 27
MALDI-TOF mass spectrometry
    175
mannose 6-phosphate receptors
    170
mannose binding protein 170
map-out putative interaction sites
    103
mapping of putative recognition
    sites 103
medicinal chemistry 76
medicinal chemistry programs 64

membrane anchor   48
micropolarity   76
MIMUMBA   112
molecular comparison   115
molecular dynamics calculations
   33
molecular ecosystem   214, 234
molecular recognition   53, 81, 97,
   115
multiple conformers   108, 112
multivalency   171, 196, 205

N-(phosphonomethyl)glycine   35
N-Cbz-l-Asp   58
N-Cbz-l-Glu   58
naphthyridine   57
netropsin   152
neuroreceptors   54
new computational tools in drug design   98
nuclear Overhauser effect   61
nuclease resistant aptamers   140
nucleic acid/ligand complexes   136
nucleotide binding region   31

oligosaccharides   184, 193
oligosaccharide libraries   170
oligosaccharide mimetics   171
optimization in Fourier space   108

π-allylpalladium complexes   10
packing in crystalline benzene   99
patterns and motifs of interaction geometries   102
PEGA   172
pentaglycyl bridges   42
peptide   214, 219, 226
peptide nucleic acids   152
peptidoglycan   42
peptidoglycan pentapeptide   47
pharmacophore   103
phosphodiester groups   65

phosphoenolpyruvate   35
photolytic release   180
physicochemical properties   108
POEPOP   172
polymerase chain reaction   135
porcine liver
– receptors   175
prion specific RNA aptamers   138
pseudoisocytosine   159
purine   120, 123, 125
pyridine   57
pyrimidine   123, 125

quaternary ammonium groups   101

R-serine   12
receptor preorganization   61
receptors   98
recognition patterns   115
REDOR   27
REDOR dephasing   32, 44
REDOR difference spectrum   38
REDOR-determined distances   33
relevant binding-site geometries
   109
ReliBase   99
ribozymes   119, 143

sample rotation   26
scrapie   140
secondary electrostatic H-bonding
   interactions   61
secondary electrostatic interactions   57
selectin   170
sequence discrimination   159
shape complementarity   97
shikimate-3-phosphate   35
sialyl-Le$_x$ 171
site-directed mutagenesis   40
solid phase libraries   172
solid-state NMR   26

spatial similarity 107
Spiegelmer 141
spiro-alkaloid nitramine 11
spirobifluorene cleft 58
split and combine synthesis 171
Staphylococcus aureus 42
steroid binding selectivities 69
steroid complexes 54
steroid solubilization 54
steroids 66, 72
Stille coupling 66
strand displacement 157
structural superposition 115
structurally well-defined pattern 97
structure-based de novo design 54
substrates 98
superimpose 107
superposition 103, 112
supramolecular synthons 98, 102, 115
supramolecule 98

Suzuki cross-coupling reaction 60
syn conformation 122
synthetic receptors 53

T-shaped interactions 62
T-shaped, edge-to-face arrays 99
template 221, 223, 225, 231
ternary complex 35
testosterone 71, 72
thiouracil 164
thymine recognition 161
torsional fragments 109
torsional libraries 112
triplex 155

van der Waals dispersion interactions 63
vancomycin 47
vancomycin complexes 47
vigabatrin 12
vinylglycinol 12

Ernst Schering Research Foundation Workshop

Editors: Günter Stock
         Ursula-F. Habenicht

Vol. 1 *(1991)*: Bioscience ⇋ Society – Workshop Report
Editors: D. J. Roy, B. E. Wynne, R. W. Old

Vol. 2 *(1991)*: Round Table Discussion on Bioscience ⇋ Society
Editor: J. J. Cherfas

Vol. 3 *(1991)*: Excitatory Amino Acids and Second Messenger Systems
Editors: V. I. Teichberg, L. Turski

Vol. 4 *(1992)*: Spermatogenesis – Fertilization – Contraception
Editors: E. Nieschlag, U.-F. Habenicht

Vol. 5 *(1992)*: Sex Steroids and the Cardiovascular System
Editors: P. Ramwell, G. Rubanyi, E. Schillinger

Vol. 6 *(1993)*: Transgenic Animals as Model Systems for Human Diseases
Editors: E. F. Wagner, F. Theuring

Vol. 7 *(1993)*: Basic Mechanisms Controlling Term and Preterm Birth
Editors: K. Chwalisz, R. E. Garfield

Vol. 8 *(1994)*: Health Care 2010
Editors: C. Bezold, K. Knabner

Vol. 9 *(1994)*: Sex Steroids and Bone
Editors: R. Ziegler, J. Pfeilschifter, M. Bräutigam

Vol. 10 *(1994):* Nongenotoxic Carcinogenesis
Editors: A. Cockburn, L. Smith

Vol. 11 *(1994)*: Cell Culture in Pharmaceutical Research
Editors: N. E. Fusenig, H. Graf

Vol. 12 *(1994):* Interactions Between Adjuvants, Agrochemical
and Target Organisms
Editors: P. J. Holloway, R. T. Rees, D. Stock

Vol. 13 *(1994):* Assessment of the Use of Single Cytochrome
P450 Enzymes in Drug Research
Editors: M. R. Waterman,
M. Hildebrand

Vol. 14 *(1995):* Apoptosis in Hormone-Dependent Cancers
Editors: M. Tenniswood, H. Michna

Vol. 15 *(1995):* Computer Aided Drug Design in Industrial Research
Editors: E. C. Herrmann, R. Franke

Vol. 16 (1995): Organ-Selective Actions of Steroid Hormones
Editors: D. T. Baird, G. Schütz, R. Krattenmacher

Vol. 17 (1996): Alzheimer's Disease
Editors: J.D. Turner, K. Beyreuther, F. Theuring

Vol. 18 (1997): The Endometrium as a Target for Contraception
Editors: H.M. Beier, M.J.K. Harper, K. Chwalisz

Vol. 19 (1997): EGF Receptor in Tumor Growth and Progression
Editors: R. B. Lichtner, R. N. Harkins

Vol. 20 (1997): Cellular Therapy
Editors: H. Wekerle, H. Graf, J.D. Turner

Vol. 21 (1997): Nitric Oxide, Cytochromes P 450,
and Sexual Steroid Hormones
Editors: J.R. Lancaster, J.F. Parkinson

Vol. 22 (1997): Impact of Molecular Biology
and New Technical Developments in Diagnostic Imaging
Editors: W. Semmler, M. Schwaiger

Vol. 23 (1998): Excitatory Amino Acids
Editors: P.H. Seeburg, I. Bresink, L. Turski

Vol. 24 (1998): Molecular Basis of Sex Hormone Receptor Function
Editors: H. Gronemeyer, U. Fuhrmann, K. Parczyk

Vol. 25 (1998): Novel Approaches to Treatment of Osteoporosis
Editors: R.G.G. Russell, T.M. Skerry, U. Kollenkirchen

Vol. 26 (1998): Recent Trends in Molecular Recognition
Editors: F. Diederich, H. Künzer

Vol. 27 (1998): Gene Therapy
Editors: R.E. Sobol, K.J. Scanlon, E. Nestaas, T. Strohmeyer

Supplement 1 (1994): Molecular and Cellular Endocrinology of the Testis
Editors: G. Verhoeven, U.-F. Habenicht

Supplement 2 (1997): Signal Transduction in Testicular Cells
Editors: V. Hansson, F. O. Levy, K. Taskén

Supplement 3 (1998): Testicular Function:
From Gene Expression to Genetic Manipulation
Editors: M. Stefanini et al.